Prime Numbers, Friends Who Give Problems

A Trialogue with Papa Paulo

Prime Numbers, Friends Who Give Problems

A Trialogue with Papa Paulo

Paulo Ribenboim

Queen's University, Canada

World Scientific

Published by

World Scientific Publishing Co. Pte. Ltd.
5 Toh Tuck Link, Singapore 596224
USA office: 27 Warren Street, Suite 401-402, Hackensack, NJ 07601
UK office: 57 Shelton Street, Covent Garden, London WC2H 9HE

Library of Congress Cataloging-in-Publication Data
Names: Ribenboim, Paulo.
Title: Prime numbers, friends who give problems : a trialogue with Papa Paulo / by Paulo Ribenboim (Queen's University, Canada).
Description: New Jersey : World Scientific, 2016. | Includes indexes.
Identifiers: LCCN 2016020705| ISBN 9789814725804 (hardcover : alk. paper) | ISBN 9789814725811 (softcover : alk. paper)
Subjects: LCSH: Numbers, Prime.
Classification: LCC QA246 .R474 2016 | DDC 512.7/23--dc23
LC record available at https://lccn.loc.gov/2016020705

British Library Cataloguing-in-Publication Data
A catalogue record for this book is available from the British Library.

Copyright © 2017 by World Scientific Publishing Co. Pte. Ltd.

All rights reserved. This book, or parts thereof, may not be reproduced in any form or by any means, electronic or mechanical, including photocopying, recording or any information storage and retrieval system now known or to be invented, without written permission from the publisher.

For photocopying of material in this volume, please pay a copying fee through the Copyright Clearance Center, Inc., 222 Rosewood Drive, Danvers, MA 01923, USA. In this case permission to photocopy is not required from the publisher.

Typeset by Stallion Press
Email: enquiries@stallionpress.com

Qu'on ne me dise pas que je n'ai rien dit de nouveau; la disposition des matières est nouvelle. Quand on joue à la paume, c'est une même balle dont joue l'un et l'autre, mais l'un la place mieux.

Blaise Pascal

Preface

Mathematicians are sometimes intelligent, but never magicians. Their ways of thinking are natural and intuitive, contorted or opaque as they may seem. Ideas, even when simple in essence, may have to be dressed up, supported by calculations and expressed in a jargon which repulses the layman. The challenge of explaining the theories and discoveries is difficult to surmount. As a result, so many are deprived of the music of mathematics.

What you will read revolves around prime numbers. Hardly any kind of numbers is more attractive or mysterious. Not much of what follows will be supported by rigorous proof. But everything will be made credible, even irrefutable.

You opened this book. You think that you are smart enough to understand all the discussions reported here. You may be surprised.

This book should not be read by specialists, who discover much and explain little.

This book is not for anyone lacking in intelligence. You, who are smart enough and even more, curious, must be warned: This is not a novel; what you read here should generate much reflection.

More than the facts described, the ways mathematicians think should captivate you.

To read and to reflect: You must.
To consult other books: You will (feel the urge).
To reach a supreme intellectual enjoyment: I hope.

P.R.

Acknowledgments?

Do I have to thank Eric and Paulo? My brain sweats in replying to their incessant and inquisitive questions, prompted by a thirst for knowledge. A thirst for a *guaraná* would be easier to satisfy.

Do I have to thank Eric and Paulo who convinced me to write this book?

Yes, but only if my readers will feel gratified, which remains to be seen.

Acknowledgments!

Once the crime was perpetrated, it was examined by my virtual friend Reny Montandon. Chosen because he allies the best of Brazil and Switzerland, rigor and friendliness, in whatever order. Thanks!

Contents

Preface	vii
Acknowledgments? Acknowledgments!	ix
1. What are Prime Numbers?	1
2. Division is Harder than Multiplication	5
3. Another Paulo! Is a Dialogue of Three Possible?	17
4. How Natural Numbers are Made Out of Primes	21
5. Tell Me: Which is the Largest Prime?	29
6. Trying Hard to Find Primes	33
7. A Formula, A Formula, Please	39
8. Paulo Came with a Lasso	47
9. Beautiful Old Elementary Arithmetic	51

10. The Old Man Still Knows	61
11. Can You Tell Me All About Congruences?	71
12. Homework Checked	85
13. Testing for Primality and Factorization	89
14. Fermat Numbers are Friendly. Are They Primes?	93
15. This World is Perfect	105
16. Unfriendly Numbers from a Friend of Fermat's	111
17. Paying My Debt	123
18. Money and Primes	159
19. Secret Messages	175
20. New Numbers and Functions	185
21. Princeps Gauss	195
22. Gathering Forces	207
23. The After "Math" of Gauss	225
24. Primes After Dinner: Bad Dreams?	239
25. Primes in Arithmetic Progression	245

26. Selling Primes	261
27. The Great Prime Mysteries	273
28. Mysteries in Sequences: More But Not All	291
29. The End and the Beginning	311
Name Index	317
Subject Index	319

1

What are Prime Numbers?

Eric. Papa Paulo, I've heard about prime numbers. Some smart people talk about these numbers with voices full of mystery. I know nothing about them. Can you explain prime numbers to me?

Papa Paulo (*the friendly Papa Paulo*). I will be happy to do it.

I tell you, if I explain everything slowly, if you pay attention and understand all that I say, you will learn about prime numbers. Then it will be your turn to impress the people who appear to know it all. Let us agree that you'll not be shy and that you will ask all the questions which come to your mind. If you have a question, my attitude is that you want to know the answer and I will provide it with patience and clarity.

Eric. At least with you I don't have to hide my ignorance.

Papa Paulo. Ignorance calls for learning, so it is a privileged situation.

Eric. After you learn, you are no longer ignorant, and the privileged position is lost. Why learn?

Papa Paulo. It is a paradox. The more you learn, the more questions and problems appear requiring answers and solutions. If we continue our dialogue for a while, you'll see how questions arise with any piece of knowledge acquired.

Eric. Your game between ignorance and knowledge is nothing so new. What I want to know is: What is a prime number?

Papa Paulo took an almost solemn air and said in an irrefutable voice:

Papa Paulo. A prime number is a natural number greater than 1 whose only factors are 1 and the number itself.

Eric. You promised to explain everything. What is a natural number?

Papa Paulo. The natural numbers are $0, 1, 2, 3, 4, \ldots, 1093, \ldots, 301258, \ldots$.

Eric. And what are the *factors of a number*?

Papa Paulo. I will give an example and you will understand right away: The factors of 10 are 1, 2, 5 and 10. These factors are also called *divisors*. Another example: the factors or divisors of 12 are 1, 2, 3, 4, 6 and 12. You see that 12 divided by 5 is more than 2 and less than 3, so 12 divided by 5 is not an integer. Therefore 5 is not a divisor (= factor) of 12.

Eric. I understand what you said. I can see that every natural number n greater than 1 has the divisors 1 and n, and sometimes has other divisors.

Papa Paulo. You are right — 1 and n are called the *trivial divisors* (= factors) of n. As I said, the number n greater than 1 is called a prime number if it has only the trivial divisors.

Eric. If one thinks just a little, making simple mental calculations, it is easy to see that 2, 3, 5, 7, 11, 13, 17, 19 are prime numbers.

Papa Paulo. Yes, Eric. Let us see how one has to proceed to find out if 19 is a prime number.

- Does 2 divide 19? No, because $9 < \frac{19}{2} < 10$, so $\frac{19}{2}$ is not an integer.
- Does 3 divide 19? No, because $6 < \frac{19}{3} < 7$.
- Does 4 divide 19? No, because $4 < \frac{19}{4} < 5$.

Continuing in this way, we find no factor of 19 except 1 and 19. So 19 is a prime number.

Eric. I see that to find if a natural number is a prime it may be necessary to test many divisors.

Papa Paulo. This is certainly a method which may be lengthy to perform for large numbers. What is important is that for every natural number $n > 1$, it is possible to decide if n is a prime or not with a finite number of operations. Much of the study of primes concerns the search for methods to decide if a given number is prime.

After these remarks (which seemed a bit mysterious to Eric), Papa Paulo said:

Papa Paulo. Let me tell a story which — believe me — is true. I had a friend who was a Nobel Prize Winner. Despite his advanced age, he continued working many hours each day in his laboratory. After his daily unrelenting concentration, my friend (who I will refer by the initials of his name, A. G.) was losing sleep. His solution? He told me that contrary to ordinary people he, a Nobel Prize Winner, would not count sheep. Instead, he would count prime numbers. It was pretty fast up to 100 because he knew them all: 2, 3, 5, 7, 11, 13, 17, 19, 23, 29, 31, 37, 41, 43, 47, 53, 59, 61, 67, 71, 73, 79, 83, 89, 97. Going beyond 100 became tiring; he confessed that he always fell asleep before reaching 200.

Eric, it is time to go to sleep. Try to reach 200!

Eric. I will not be able to concentrate if you don't tell me the name of A. G., your Nobel Prize Winner.

Papa Paulo. I pass the secret to you: APPOLONIUS GSCHWENDTNER.

Eric. With this name he could not become a sumo champion, so he had to be a Nobel Prize Winner.

2

Division is Harder than Multiplication

The next day, Papa Paulo began:

Papa Paulo. I can see you are anxious to know more about prime numbers, and I will satisfy you soon. But it is convenient to first discuss multiplication and division, pointing out some simple facts that will be needed.

Papa Paulo paused to give a nice effect to what he was going to say:

Papa Paulo. Mathematics, like many sciences, is like a building, a store supported by others. A monumental building. If stones are missing, the building may collapse. If one wants to learn mathematics, it is necessary to because familiar with the stones of the base, and then with the stones just above, and so on. Each stone in equilibrium on top of others. Today we shall get acquainted with two "stones": multiplication and division.

Now Eric, who understood what I said, but wanted to provoke me, said:

Eric. Little stones make up the abacus that was widely used in olden times, even today; to add, subtract, multiply and divide.

I cut him short and said:

Papa Paulo. I spoke about stones in a figurative way.

Multiplication

If n is a natural number, the numbers $n \times 0, n \times 1, n \times 2, n \times 3, \ldots$, are called the *multiples* of n. 0 has only the multiple 0. The multiples of 1 are all natural numbers. The multiples of 2 are all *even* natural numbers. The natural numbers which are not even are called *odd*.

Eric. Everyone knows this.

Papa Paulo. Now I will tell you about square numbers, cube numbers, units, and powers. You know that if you have a square with sides measuring 3 then the area measures $9 = 3 \times 3$. For this reason we say that 9 is the square of 3 and we write $9 = 3^2$. Here are the squares of some other numbers:

$0^2 = 0$, $1^2 = 1$, $2^2 = 4$, $3^2 = 9$, $4^2 = 16, \ldots, 100^2 = 10\,000$, $10\,000^2 = 100\,000\,000$, etc.

Eric. I can guess what you are going to say about cube numbers. If you have a cube with sides measuring 3 units then the volume of the cube is $3 \times 3 \times 3 = 27$. So we say that 27 is the cube of 3 and write $27 = 3^3$.

Papa Paulo. You've got it right. Here are the cubes of some numbers:

$0^3 = 0$, $1^3 = 1$, $2^3 = 8$, $3^3 = 27$, $4^3 = 64, \ldots, 100^3 = 1\,000\,000$, $1\,000\,000^3 = 1\,000\,000\,000\,000\,000\,000$. You see, for squares you doubled the number of zeros, for cubes you need 3 times the number of zeros.

Eric. Of course you may also multiply any number with itself 4 times, say $2 \times 2 \times 2 \times 2$.

Papa Paulo. $2 \times 2 \times 2 \times 2$ is also written 2^4 and is called the fourth power of 2. In the same way we define the fifth power 2^5, the sixth power 2^6, and so on. To get a feeling of how these powers become bigger and bigger, here are two examples

$2^2 = 4$, $2^3 = 8$, $2^4 = 16$, $2^5 = 32$, $2^6 = 64$, $2^7 = 128$, $2^8 = 256$, $2^9 = 512$, $2^{10} = 1024$, and so on.

$3^2 = 9$, $3^3 = 27$, $3^4 = 81$, $3^5 = 243$, $3^6 = 729$, and so on.

Eric. The powers of 10 are the easiest to calculate. 10^7 is just 1 followed by 7 zeros.

Papa Paulo. People are often surprised by how fast powers of numbers greater than 1 become large. Let me tell you a famous story. One day, the Sultan of a medieval country received the gift of a chess game which had just been invented by one of his subjects. The game was played in front of the Sultan, who loved it so much that he wanted to give a substantial reward to the inventor.

The subject: Your Highness is very kind in wanting to compensate your humble subject for such a modest and insignificant gift.

Upon the insistence of the Sultan, the humble inventor said:

The subject: Please give me 1 grain of wheat for the first square of the chessboard, 2 grains for the second square. For the third square, give me 4 grains of wheat and so on. For each square, double the grain that was given for the previous square.

The Sultan (to his Vizier): Do as my subject has asked.

Papa Paulo. Eric, I will interrupt the story; I want you to calculate how many grains of wheat the Sultan should give to the humble — but smart — inventor of the chess game. You will realize that the Vizier had to tell the Sultan that in the whole world there was not enough wheat to make the requested payment.

Eric. So how did the story end?

Papa Paulo. I guess that the subject was invited to take care of the finances of the Sultan. Some wheat was left in the world because till today we still have bread.

Division

Papa Paulo. Sometimes the result of a division is a natural number. For example, 12 divided by 3 is equal to 4. We write: $\frac{12}{3} = 4$, or also $12 = 4 \times 3$, and we say that 12 is a multiple of 3, but also a multiple of 4; we may also say that 3 and 4 divide 12, or that 3 and 4 are divisors, as well as factors, of 12. All of these expressions are interchangeable.

Other times, the result of a division is not a natural number. For example: $\frac{13}{3} = 4 + \frac{1}{3}$. This may also be written as $13 = 4 \times 3 + 1$. We say that 4 is the quotient of the division and 1 is the remainder of the division of 13 by 3.

Try the following divisions:

    100 divided by 17
    1 000 divided by 171
    1 000 000 divided by 1717
    91 divided by 7

After you finish write the result in the way I did:
    (quotient) $\times$ 17 + (remainder)
and so on.

A few minutes later, Eric came with the answers. I asked him to check by doing the multiplication and adding the remainder to find the given number. This was easier to do, because division is harder than multiplication.

The Euclidean Division Theorem

Papa Paulo. I'll give a theorem about *any* natural numbers; for this reason I shall use letters to represent numbers. Did I already explain the signs $<$ and $>$?

Eric. No.

Paulo. If I write $3 < 5$ this is read: "3 is smaller than 5".

In the same way, $5 > 3$ is read "5 is bigger than 3". So
$$0 < 1 < 2 < 3 < 4 < 5 < \ldots.$$

Any number greater then 0 is called a *positive number*, so they are $1, 2, 3, 4, \ldots$.

We also have
$$0 > -1 > -2 > -3 > -4 > -5 > \ldots.$$
The numbers $-1, -2, -3, -4, -5, \ldots$ are the *negative numbers*.

Eric. I am thrilled to see a theorem. For me it is the first time. Will I understand it, or are theorems just for special brains? Tell me what is a theorem?

Papa Paulo. A theorem is a mathematical statement which is true. The truth is proven by logically irrefutable arguments.

Papa Paulo. You will understand the Euclidean division theorem because you have already practiced some divisions. Here is the theorem:

Euclidean Division Theorem. *Let a and d be positive natural numbers. Then,*

(1) *There exist numbers q and r such that $r < d$ and $a = q \times d + r$.*
(2) *There is only one number q and only one number r with the properties indicated above.*

Eric. You are telling two things. The first one causes me no trouble. If I say $a = 237$ and $d = 13$ then by division I get $237 = 18 \times 13 + 3$, so I can take $q = 18$ and $r = 3$. I'm happy with part (1).

Papa Paulo. I'm glad you are happy, but what you showed with the numbers $a = 237$ and $d = 13$ is not a proof which is good for all numbers a and d, whatever be the choices of a and d.

Here is the proof.

The numbers
$$d = 1 \times d < 2 \times d < 3 \times d < \ldots$$
are increasing and increasing. So there is a number q such that a is still greater or equal to $q \times d$, but already smaller than $(q+1) \times d$. So $q \times d \leq a < (q+1) \times d$. Subtracting $q \times d$ gives
$$0 = q \times d - q \times d \leq a - q \times d < (q+1) \times d - q \times d = d.$$
Let r be the number $r = a - q \times d$, so $0 \leq r < D$ and $a = q \times d + r$.

This is the proof of part (1).

Eric. It is so simple, I cannot believe it, but I am still puzzled by part (2).

Papa Paulo. This is just as easy to prove. We already know that $a = q \times d + r$ with $0 \leq r < d$. We ask ourselves: suppose that q_1 and r_1 are numbers such that $0 \leq r_1 < d$ and $a = q_1 \times d + r_1$. We want to show that q_1 has to be equal to q and r_1 has to be equal to r.

We have $a = q \times d + r = q_1 \times d + r_1$.

Two cases are possible:

Case 1: Assume that $q = q_1$. Then $q \times d = q_1 \times d$ hence r and r_1 must be equal. We are happy in Case 1.

Case 2: q_1 is not equal to q (this is written as $q_1 \neq q$). One of the two numbers is bigger than the other, say

$$q_1 > q \text{ so } q_1 - q \geq 1.$$

The proof is similar if $q > q_1$.

Now

$$d = 1 \times d \leq (q_1 - q) \times d = q_1 \times d - q \times d$$
$$= (a - q \times d) - (a - q_1 \times d) = r - r_1.$$

Is this possible? No, because $r - r_1 \leq r < d$. Impossible. So Case 2 cannot happen and this is the end of the proof of part (2).

q.e.d.

Usually you'll see three letters q.e.d. at the end of a proof. They are the initials of the Latin expression "*quod erat demonstrandum*", which means "what was required to prove". Now people have forgotten Latin and one uses marks like ■ or □ at the end of the proof.

Eric. Now I understand why proofs are done with letters which may represent *any* number. If I choose special numbers, I only know that the theorem is true for the chosen numbers.

Papa Paulo. We may choose any letter, capital letter or lower case letter. It is also common to use Greek letters α, β, γ, δ, ... No doubt it may impress readers who are not Greek.

Eric still had something to ask.

Eric. Euclidean comes from Euclid. Who was this person?

Papa Paulo. He was a famous mathematician from Ancient Greece. Later I'll give you a very short biography of Euclid.

Eric. I know that he is dead, because there is a Euclid Street in town.

A sad shadow passed over Eric's eyes. He said:

Eric. I hope I'll never see a Papa Paulo Street anywhere.

The Greatest Common Divisor

Papa Paulo. If we take two positive integers a and b, they will have the common divisor 1. Since all common divisors of a and b are less or equal than both a and b, there exists the *greatest common divisor* of a and b. This number is designated by $\gcd(a, b)$.

Eric. I'd like to see an example with numbers to learn how to calculate the greatest common divisor.

Papa Paulo. — **Problem**: To find the greatest common divisor of 186 and 100.

Solution: Divide 186 by 100
$$186 = 1 \times 100 + 86.$$
Divide 100 by the remainder 86
$$100 = 1 \times 86 + 14.$$
Divide 86 by the remainder 14
$$86 = 6 \times 14 + 2.$$
Divide 14 by the remainder 2
$$14 = 7 \times 2 + 0$$
Now stop and say
$$\gcd(186, 100) = 2.$$

Check that it is true. Of course 2 divides 186 and 100. If a number $d \geq 1$ divides 186 and 100 then it divides $86 = 186 - 100$
hence also $14 = 100 - 1 \times 86$
and also $2 = 86 - 6 \times 14$.

So 2 is indeed the greatest common divisor of 186 and 100. Note that I did not have to find the factors of 186 and of 100.

Eric. A very nice procedure. It was not necessary to make the list of the factors of 186 or one for 100 and then compare the two lists to find the greatest number present in both lists.

Papa Paulo continued:

Papa Paulo. I still want to make a simple but useful remark:

Going backwards:

$$2 = 86 - 7 \times 14$$
$$= 86 - 7 \times (100 - 86)$$
$$= 8 \times 86 - 7 \times 100$$
$$= 8 \times (186 - 100) - 7 \times 100$$
$$= 8 \times 186 - 15 \times 100.$$

So I could find two numbers 8 and -15 and $2 = \gcd(186, 100) = 8 \times$ (the larger number 186) $-15 \times$ (the smaller number 100). You may wonder why I stress this expression of the gcd — it will be useful in many proofs.

Eric. Now is my turn to ask for a theorem, using letters instead of numbers.

Papa Paulo. I will state the theorem.

Greatest Common Divisor Theorem: *Let a_1, a_2 be given numbers, with $a_1 > a_2 > 0$, let d denote the $\gcd(a_1, a_2)$. Then d is obtained as follows, by successive Euclidean divisions*

$$a_1 = q_1 \times a_2 + a_3$$
$$a_2 = q_2 \times a_3 + a_4$$

and so on, until after a number of repetitions one reaches

$$a_{n-1} = q_{n-1} \times a_n + 0.$$

Then $d = a_n$. Moreover there exist numbers s and t, different from zero, one of which is negative, such that $d = sa_1 + ta_2$. If $a_1 = a_2$ it is obvious that $dA_1 = 1d_2 = 2a - 1a_2$.

I could give the proof, but it is what I just did in the numerical example, so it is unnecessary to explain it. Instead, I propose that you find $\gcd(204, 357)$. For a smart person, one worked example should be sufficient to learn the procedure.

After taking a breath, Eric asked (already guessing what the answer would be):

Eric. Can one find the greatest common divisor of three numbers with a repetition of what was done for two numbers?

Papa Paulo. Let $a_1 > a_2 > a_3 > 0$ be given numbers. Let $d_1 = \gcd(a_1, a_2)$ be calculated as it was already explained. If $d_1 = a_3$ let $d = d_1$. If $d_1 \neq a_3$ let $d = \gcd(d_1, a_3)$. Then $d = \gcd(a_1, a_2, a_3) =$ greatest common divisor of $a_1, a_2\ a_3$. And again there exist integers s_1, s_2, s_3 not equal to 0 such that $d = s_1 a_1 + s_2 a_2 + s_3 a_3$.

Eric. I will not ask for the proof, since I anticipate that it will not be difficult. And I am certain that a similar theorem exists for more than three numbers.

Papa Paulo. We wrap up this discussion with one more exercise for you: Find the $\gcd(91, 343, 1575)$.

For more than two positive numbers, we have
$$\gcd(a_1, a_2, a_3) = \gcd(\gcd(a_1, a_2), a_3)$$
and the result is the same for any ordering of the numbers. Also
$$\gcd(a_1, a_2, a_3, a_4) = \gcd(\gcd(a_1, a_1, a_3), a_4)$$
but is also equal to $\gcd(\gcd(a_1, a_2), \gcd(a_3, a_4))$, etc.

Moreover, we may write
$$\gcd(a_1, a_2, a_3, a_4) = s_1 a_1 + s_2 a_2 + s_3 a_3 + s_4 a_4$$
where the numbers s_1, s_2, s_3, s_4 are not 0, some of them being negative.

The Least Common Multiple

Papa Paulo. If we have two positive numbers, say 10 and 18, they have a common multiple which is 10×18. Hence among the common multiples there must be the smallest, which is called the *least common multiple* of 10 and 18 and denoted by $\mathrm{lcm}(10, 18)$. To calculate it, one writes the list of multiples of 10, the list of multiples of 18 and determines the smallest number present in both lists.
Multiples of 18: 18   36   54   72   90   108 ...
Multiples of 10: 10   20   30   40   50   60   70   80   90   100 ...

So 90 is the least common multiple of 10 and 18. The abbreviation is $90 = \mathrm{lcm}(10, 18)$.

Eric. I suppose you can do the same for several numbers $a, b, c, d, \ldots$.

Papa Paulo. Indeed, I'll let you calculate $\mathrm{lcm}(10, 18, 25)$.

First you do — we just did it — $\mathrm{lcm}(10, 18) = 90$, then you do $\mathrm{lcm}(90, 25)$. Go ahead and do it. It doesn't matter in which order you work with the given numbers 10, 18, 25. You may first do $\mathrm{lcm}(10, 25)$ and then the least common multiple of this number and 18. All this is so easy to prove that you may figure it out by yourself.

Papa Paulo. Let us see if we can discover this proof together.

I give you the positive numbers a and b, let $\ell = \mathrm{lcm}(a, b)$. Of course each multiple of ℓ is a multiple of a and of b. Now I assert: if c is a common multiple of a and b, then c is a multiple of ℓ. Can we prove this statement?

Eric. Let me try. I have to compare c and ℓ. I must show that c is a multiple of ℓ. You explained Euclidean division. So I think a good idea is to divide c by ℓ and show that the remainder is 0: $c = q\ell + r$ with $0 \leq r < \ell$. I look at this and think: ℓ is a multiple of a and of b, c is a multiple of a and of b. Hence r is a multiple of a and of b.

Now I have to say: if $r \neq 0$, since $r < \ell = \mathrm{lcm}(a, b)$ I would have a contradiction. So $r = 0$. Great. I proved it and I can use the three Latin letters, q.e.d.

Papa Paulo. Bravo! You are smart!

Tired of greatest common divisors and least common multiples, Eric inquired:

Eric. Papa Paulo, when are you going to explain prime numbers? All you did up to now is nice but (raising his voice) I WANT PRIME NUMBERS!

To which Papa Paulo used a very convincing word:

Papa Paulo. Mañana.

In the afternoon, Papa Paulo prepared some notes about Euclid.

Notes About Euclid
(circa 300 BCE–circa 279 BCE)

Today, people have a "curriculum vitae", or c.v., where it is written when and where the individual was born, name of father, name of mother, including her maiden name, the name of the elementary schools, high schools and universities attended. Masters degree, Ph.D. and (very important) athletic awards. If appropriate, titles of prominent papers are also mentioned.

It was not so in the time of Euclid. We know he lived in Greece around 300 BCE, and may have died around 279 BCE, perhaps in Alexandria, then a Greek city, now in Egypt. We don't know where he was born, nothing about his personal life except that he was influenced by the teachings of Plato. We also know that he had a mathematical school in Alexandria which was directed after his death by his disciple Appolonius. It is unfortunate that we know so little about his life, but we know the most important part: His work. It included "*The Elements*", divided into 13 parts. In this book, Euclid used the "axiomatic method". Some statements about the concepts of the theory which were "self-evident" were taken as axioms. By logical deduction, performed with (almost) unfailing rigor, Euclid proved his theorems. This method became fundamental for the development of mathematics and it is in widespread use till today. Parts 1 to 6 of "*The Elements*" are about plane geometry. Parts 7 to 9 concern numbers and are of special interest for our present discussions. Book 10 is about irrational numbers. The other parts are concerned with solid geometry and contain, among other things, a description of the five regular convex polyhedra described by Plato, namely the tetrahedron, the cube, the octahedron, the dodecahedron and the icosahedron.

These books were translated into Latin and Arabic and were carefully studied by the mathematicians. Later translations into the major European languages perpetuated the influence of this book and its method of mathematical exposition.

3

Another Paulo! Is a Dialogue of Three Possible?

Eric brought his friend Paulo, who also wanted to learn about prime numbers. With two Paulos it will be confusing, but of course I am the Papa Paulo, so at worst they will call me P.P. — but I don't like it. Try it and you'll know why.

For the new Paulo I felt that I should repeat the definition of a prime number:

P.P. An integer p, greater than 1, which does not have factors (except the trivial factors 1 and p) is called a *prime number*. Often, we just say "a *prime*". Eric, we have already encountered some primes out of our calculations to find factors. So 2, 3, 5, 7 are the prime numbers less than 10. With some trouble you found that 97 is a prime and with more calculations, we came to the conclusion that 911 is also a prime.

Eric. It is very easy to understand what a prime number is, but why would anyone be interested in primes?

P.P. A prime number is like an atom. The word "atom" comes from Greek and means "it cannot be broken into parts". You know that molecules are made up from atoms. One of these days I'll tell you that all natural numbers are "made up from primes".

Eric. We had already found many numbers which are primes. For testing we had to perform divisions to find factors, if any existed. No big deal for small numbers, like 97. But it was much more work for 911. You know Papa Paulo, I am LAZY! Isn't there a better way to find if a number is prime without performing so many divisions?

P.P. Laziness is a great human quality. It is the mother of invention. Men discovered that the shortest distance between two points is to follow the straight line joining the points. Men invented the wheel to support vehicles and transport himself or merchandise. And elevators to avoid climbing stairs.

Papa Paulo continued:

The Sieve

P.P. There is a simple method to find primes without performing any division; I will tell you how to do it. We shall find all primes up to 100. Just follow my instructions. We already know that 2 is a prime. In fact it is the only even prime.

(1) Write 10 rows of odd numbers, like

    1   3   5   7   9

  11 13 15 17 19   etc.

(2) Cross out 1.

(3) Do not cross out 3, jump 3 numbers
          and cross out 9, jump 3 numbers
          and cross out 15, and continue this way.

(4) The smallest number not yet crossed out is 5. Do not cross it out but jump 5 numbers. Do not cross out 15, which was already not crossed. Jump 5 numbers and cross out 25. Repeat this procedure.

(5) The smallest number not yet crossed out is 7 — keep it. Then jump 7 numbers. 21 is already crossed out. Jump 7 numbers to 35, already crossed out, jump again 7 numbers, cross out 49, and repeat the procedure.

(6) Stop!

Papa Paulo continued the explanation:

P.P. Numbers crossed out are multiples of smaller numbers (not equal to 1), so they are not primes. For the same reason, numbers not crossed out are primes.

Paulo. Why did you stop after jumps of 7?

P.P. After 7, I had to keep 11 and jump 11 numbers, so I would stop on 11×3, 11×5, 11×7, already crossed out when jumping by 3, by 5, and by 7. If a jump of 11 numbers is done, one would reach $11 \times 11 = 121 > 100$, which is beyond the range of the table.

In summary, the primes less than 100 are 2 and all the numbers that were not crossed out: 2, 3, 5, 7, 11, 13, 17, 19, 23, 29, 31, 37, 41, 43, 47, 53, 59, 61, 67, 71, 73, 79, 83, 89, 97.

Here is the table obtained:

~~1~~	2	3	~~4~~	5	~~6~~	7	~~8~~	~~9~~	~~10~~
11	~~12~~	13	~~14~~	~~15~~	~~16~~	17	~~18~~	19	~~20~~
~~21~~	~~22~~	23	~~24~~	~~25~~	~~26~~	~~27~~	~~28~~	29	~~30~~
31	~~32~~	~~33~~	~~34~~	~~35~~	~~36~~	37	~~38~~	~~39~~	~~40~~
41	~~42~~	43	~~44~~	~~45~~	~~46~~	47	~~48~~	~~49~~	~~50~~
~~51~~	~~52~~	53	~~54~~	~~55~~	~~56~~	~~57~~	~~58~~	59	~~60~~
61	~~62~~	~~63~~	~~64~~	~~65~~	~~66~~	67	~~68~~	~~69~~	~~70~~
71	~~72~~	73	~~74~~	~~75~~	~~76~~	~~77~~	~~78~~	79	~~80~~
~~81~~	~~82~~	83	~~84~~	~~85~~	~~86~~	~~87~~	~~88~~	89	~~90~~
~~91~~	~~92~~	~~93~~	~~94~~	~~95~~	~~96~~	97	~~98~~	~~99~~	~~100~~

The primes are the numbers which were not crossed out.

Eric. It is obvious that if we use a computer, we could find all the primes up to 1 000 000. The computer deals with all the odd numbers less than 1 000 000; by means of a simple program it will determine prime numbers up to 1 000 000.

Paulo. This is a great procedure to find primes without ever dividing. Papa Paulo, did you invent this method to find primes?

P.P. It was invented in Ancient Greece, and is called the *Eratosthenes Sieve*.

Eric kept thinking and realized:

Eric. To find the primes up to 1 000 000 and write them on paper, there will be 100 000 rows; in a notebook with 25 rows per page, I need 4000 pages, so I need 40 notebooks of 100 pages each. To have the primes up to 1 000 000 000 I will need 40 000 notebooks.

He concluded:

Eric. Only computers can calculate this and store such large lists. Have primes bigger them 1 000 000 000 been found?

P.P. Yes. I'll discuss this question extensively. Many people try to be famous by finding *one* prime which is larger than any other known prime at that particular time. These people are only famous until the moment that their records are broken.

I know that there is a very popular book about prime number records, but I forget the author's name. For me, these records are just like swimming records. They have only a temporary value.

Notes About Eratosthenes (circa 276 BCE – circa 196 BCE)

We know his approximate date of birth, 276 BCE, and of his death, 196 BCE. He was born in Cyrene (now in Libya) and spent his career in Alexandria, where he died. Eratosthenes was a very respected scholar, who occupied for many years the position of Chief Librarian at Alexandria. The areas of activity of Eratosthenes were geography, mathematics, chronology, philosophy, and grammar. He made outstanding contributions in geography. In mathematics, he dealt with the problem of duplication of the cube and invented the sieve which perpetuated his name in number theory.

4

How Natural Numbers are Made Out of Primes

The day came to fulfill my promise to show Eric and Paulo (the other Paulo) how natural numbers are made out of primes. The fundamental theorem was proven by Euclid.

Paulo. I don't know why, but I say "... by a Greek." I have not yet learned to spell that hard name Erat... So I say "Euclid".

P.P. And you are right. The theorem has two parts — the first one is easy, but for the proof of the second part, some preparation will be needed.

Eric. Please start with the easy one.

P.P. Part 1 (the easy part) of Euclid's Theorem: Every natural number greater than 1 is the product of primes. Note that 0 and 1 are excluded. The proof is interesting — though easy — and I'll ask your help, Paulo. I don't know if you "believe" that Part 1 is true, but "make believe" that you "don't believe". A lot of beliefs at this point, but soon there will be a logical, irrefutable proof.

Paulo obeyed and said:

Paulo. OK, Papa Paulo, I don't believe that Part 1 is true.

P.P. That means that you believe there is a natural number greater than 1 which is not a product of primes.

Paulo. Yes, I agree.

P.P. Then there is the smallest natural number, which is greater than 1, but not a product of primes.

Paulo. Yes, I agree again, but what is this number?

P.P. As a matter of fact, it is a hypothetical number, so I will call it n. My intention is to show that no such number n exists. Now I ask, could n be a prime number?

Paulo. The answer is "NO" because if n was prime then n would be the product of just one prime and our n is not a product of primes.

P.P. Very good, so what is next?

Paulo. n must have a factorization $n = m \times q$ with $1 < m < n$ and $1 < q < n$.

P.P. Great, so both m and q are products of primes, because m and q are smaller than n, and n was the smallest number not equal to a product of primes. So $n = m \times q =$ (product of primes) $\times$ (product of primes) = (product of primes).

Paulo. I know you are not a liar, but you've got an absurdity. Why?

P.P. You said that Part 1 was false, so...

Paulo concluded:

Paulo. Part 1 is true! This is the proof. Papa Paulo you are a magician. I feel as if you brought me on stage and performed your magic on me!

P.P. I showed you a common method to prove theorems:

(a) As a provocation you say that the theorem to be proven is false.
(b) The magician (the mathematician) does some work and ...
(c) Reaches an absurdity.

So the theorem has to be true.

The method is called *reduction to absurdity*, or in Latin, *Reductio ad absurdum*.

It is also used by good lawyers to destroy arguments of opponents.

How Natural Numbers are Made Out of Primes

Throughout the discussion Eric observed with pleasure that Paulo could indeed actively participate and help in reaching the absurd conclusion. Eric now asked for the more difficult Part 2.

P.P. First there must be some preparation.

Here is a definition which I am giving you for the first time: The natural numbers a and b (different from 0), are said to be *coprime* (some people say *relatively prime*) when their greatest common divisor is equal to 1.

As I said earlier, if a and q are coprime, it is possible to find integers r, s, not equal to 0, such that $1 = ra + sq$. I use this fact to show: If a and b are natural numbers, not equal to 0, if $q > 0$ is a factor of ab and q and a are coprime, then q is a factor of b.

Proof: We have $1 = ra + sq$. Multiplying with b we obtain $b = rab + sqb$. By assumption q divides ab, so q divides rab and sqb, so q divides their sum b. q.e.d.

Note that when $q = p$ is a prime, to say that p and a are coprime is just the same as saying that p does not divide a.

Paulo. I know why. If p does not divide a, the only factor of p and a must be 1, that is p and a are coprime. The converse is even easier.

Eric. So I may say: If the prime p divides ab, but p does not divide a, then p divides b.

And there is no need for a new proof.

P.P. Believe me, I can prove in the same way that if $a_1, a_2, \ldots, a_k$ are non-zero natural numbers, if p is a prime and p divides the product $a_1 \times a_2 \times \cdots \times a_k$, then p divides one of the numbers $a_1, a_2, \ldots, a_k$.

Eric. I'll try to do it at home, it will be easy. Please state and prove Part 2 of Euclid's theorem.

P.P. This is Part 2: A natural number $n > 1$ cannot be the product of primes in more than one way, except for allowing a change in the order in which the primes are multiplied. Here is the proof:

Proof: Let h, k be natural numbers. If n is the product of h primes and also the product of k primes then (product of h primes) =

(product of k primes). I wish to show that $h = k$ and the primes in the left-hand product are the same as the primes in the right-hand product. Suppose that $h \neq k$; we may then assume that $h > k$; it would be the same proof if $h < k$. Let us see what happens. Pick a prime p in the left-hand product; p divides the right-hand product, so p divides one of the primes, hence it is equal to one of the primes in the right-hand. Therefore $p \times$ (product of $h - 1$ primes) $= p \times$ (product of $k - 1$ primes). Dividing by p, we obtain (product of $h - 1$ primes) = (product of $k - 1$ primes), the same situation as before but with less primes. So we repeat the same process several times, getting that k of the primes in the left-hand side are equal to primes in the right-hand side. From $h > k$ we are left with $h - k$ primes in the left-hand side and no primes in the right-hand side, that is (product of $h - k$ primes) $= 1$. This is impossible, so $h = k$, and the proof is finished. q.e.d.

Eric. I like your proof. You were able to explain the proof in convincing terms, so I could follow the idea. You did not mess up with symbols which can sometimes make the proof seem difficult.

After a while, Eric added:

Eric. And you know? I like proofs when you repeat and repeat the same argument, getting smaller and smaller numbers until everything is obvious.

P.P. This is an important method. It is called the *method of descent*. It appears under other guises, and then it is called the *method of (mathematical) induction*. Perhaps one day I'll show you a proof by induction.

Paulo, the theorem-lover, said:

Paulo. I also understood the proof, but why is this theorem so important? What can you do with it?

P.P. If you have a problem involving natural numbers, you can consider the problem for powers of primes, solve the problem for powers of primes, put the solution for each prime together in an appropriate way, to solve the original problem.

Paulo. I don't know what you are saying. Give us a concrete example.

P.P. We examine this problem together: to find all factors of 2592.

First we shall find the factorization as a product of primes. My calculations are easy to understand:

$$\begin{array}{r|l} 2592 & 2 \\ 1296 & 2 \\ 648 & 2 \\ 324 & 2 \\ 162 & 2 \\ 81 & 3 \\ 27 & 3 \\ 9 & 3 \\ 3 & 3 \\ 1 & \end{array}$$

so $2592 = 2^5 \times 3^4$.

The factors of 2^5 are $1, 2, 2^2, 2^3, 3^4, 2^5$.
The factors of 3^4 are $1, 3, 3^2, 3^3, 3^4$.

A factor of 2^5 times a factor of 3^4 is a factor of $2^5 \times 3^4$ and vice-versa. Do you understand why "vice-versa"?

Eric. Yes, I don't even have to tell you because it is so easy with all that we already know. So $2592 = 2^5 \times 3^4$ has $6 \times 5 = 30$ factors, which are all the numbers $2^m \times 3^n$, where $m = 0, 1, 2, 3, 4$ or 5 and $n = 0, 1, 2, 3$ or 4.

Eric. What else can you do with Euclid's Theorem?

P.P. Take the number $467775 = 3^5 \times 5^2 \times 7 \times 11$. It is now very easy to find $\gcd(2592, 467775)$. I call d the gcd I wish to determine. It is a product of primes. A prime p is a factor of d exactly when it divides both numbers 2592 and 467775. Only 3 is possible. The highest power of 3 dividing both numbers is 3^4, so $d = \gcd(2592, 467775) = 3^4 = 81$.

Eric. So the rule to find $d = \gcd(a, b)$ is as follows:

(1) find the factorizations of a and of b as products of primes;

(2) pick the primes p appearing in both factorizations;
(3) take p^e for each common prime, where e is the minimum of the exponents of the prime dividing each one of the numbers;
(4) d is the product of all these prime powers p^e.

P.P. That is it.

Paulo. And how about the least common multiple? How do you get $\ell = \text{lcm}(2592, 467775)$?

P.P. Looking at the factorizations, I see right away that $\ell = 2^5 \times 3^5 \times 5^2 \times 7 \times 11$. I use all prime powers p^e, where p divides either a or b, e is the maximum of the exponents.

Paulo. You need not explain any longer. It is quite clear.

Eric. OK! OK! All this is easy, but you must know the factorizations of the given numbers as products of powers of primes. For large numbers, we already know that this may take a long time to do.

P.P. As a matter of fact, it is possible to avoid the factorization. This is how it goes, for any two numbers a, b greater than 0:

(1) make the product ab
(2) find $d = \gcd(a, b)$ by successive Euclidean divisions; no factorization is needed
(3) $\ell = \text{lcm}(a, b) = \frac{ab}{d}$; so no factorization is needed.

Eric. But the proof that $\ell d = ab$ is needed.

P.P. In this proof I apply the theorem of factorization. Let p be any prime dividing either a or b. Let $p^e =$ power of p appearing in the factorization of a, $p^f =$ power of p appearing in the factorization of b. If p does not divide a, put $e = 0$. If p does not divide b, put $f = 0$. But both $e = 0$ and $f = 0$ are not allowed.

Then p^{e+f} is the power of p in the factorization of ab.

Now I look at the product $d\ell$. Assume that $e \geq f$ (it would be similar if $f \geq e$). Then p^f is the power of p in the factorization of d, p^e is the power of p in the factorization of ℓ.

Then p^{e+f} is the power of p in the factorization of $d\ell$.

Compare $d\ell$ and ab. All primes p appear with the same exponent $e + f$ in $d\ell$ and ab. Hence, $d\ell = ab$. q.e.d.

Eric. Good! The theorem was used to show how factorization could be avoided in the calculations; kind of funny. Despite this getting around factorization, in many situations, you must know the factors. For large numbers there could be trouble.

P.P. Large numbers, large primes, large factors.... Why aren't all numbers small? Tomorrow the theme will be: Is there a largest prime?

5

Tell Me: Which is the Largest Prime?

Eric and Paulo returned to resume our conversations, this time about the "hot question".

Eric. Which is the largest prime?

No answer was forthcoming; Papa Paulo remained silent. Eric thought: Papa Paulo has not answered. Either he does not know the answer or there is no largest prime. But, Papa Paulo knows everything — he has an encyclopedia inside his head — so I suspect that there is no largest prime.

Eric. I know that there is no largest prime. I guessed it because in the Eratosthenes Sieve, primes kept appearing and that 911 is a prime. I figured out that there is no reason that this should stop.

P.P. You are right; you know how to make good guesses. But your reasons cannot be accepted. There is nothing wrong in making guesses, in fact it is a practice I very much encourage. This is what you should be doing:

(a) You guessed that there is no largest prime.
(b) You extended the sieve for larger and larger numbers. As you keep finding primes, you feel forced to ...
(c) ... prove that there is no largest prime.

Eric. I thought I could show that there was no largest natural number by adding 1 to find a bigger one. If I do this with a prime (not equal to 2) I get an even number, not a prime. So I tried adding 2:

$3 + 2 = 5$ prime

$5 + 2 = 7$ prime

$7 + 2 = 9$, not prime — so the idea was wrong! I don't know what to do.

P.P. No wonder. Again, it was Euclid — the same Euclid who did the proof for the first time. After Euclid, many different proofs were found.

In that book of prime number records which I mentioned earlier — now I know the name of the author, OLUAP MIOBNEBIR, who is rather obscure and cannot even be found on the Internet — there are many proofs of Euclid's Theorem.

Paulo. Please, tell us exactly what is Euclid's Theorem?

P.P. Euclid's Theorem. *There exist infinitely many primes.*

I will give the simplest proof I know:

Proof: If there is a finite number of primes, then there is a largest prime, which I call N. I will show that there is a prime even bigger than N. For this purpose, consider the number $A = 1 +$ (product of all the integers $1, 2, 3, \ldots,$ up to N). As I showed earlier, A is divisible by some prime number, which I call p. I'll prove that $p > N$. We assume the contrary, that is $p \leq N$ and show that this is impossible. Why? If $p \leq N$ then p divides the product of the numbers $1, 2, \ldots, N$. Since p also divides A, then p divides $A -$ (product of $1, 2, \ldots, N) = 1$, which is impossible. Thus $p > N$.

q.e.d.

Eric and Paulo listened intently, but seemed a bit puzzled.

Paulo. Your way of proving with reduction to an absurdity does not satisfy me. I thought everything mathematicians say is clear and not absurd.

P.P. Think about what was clear. If it is assumed that the theorem is false, the reasoning led to an absurdity. In mathematics no absurdity is tolerated. So the theorem had to be true.

Paulo. It is the second time you have used such a method.

Eric. I liked that to find there is a prime $p > N$ you don't need to calculate (product of $1, 2, \ldots, N$). Your arguments guarantee that such a prime p exists.

P.P. Yes, but it does not tell how to effectively find this prime. We just know that it is there, bigger than N, and this is good enough for the proof.

We stopped, knowing that we had just learned an important theorem. The world would be very different with only a finite number of primes.

Eric. I'd like to know the names of heroes. Who did this simple proof of Euclid's Theorem?

P.P. His name is STIELTJES. Let us meet again in three days.

Eric needed to say what bothered him:

Eric. Papa Paulo, I want to see how it is possible to obtain a new prime from the primes already known, not just that a new prime exists, without telling which is this new prime.

Notes About Thomas Jan Stieltjes (1856–1894)

He was Dutch, born in 1856 and lived only 38 years. In his short life, Stieltjes wrote papers of the greatest importance. His work on the analytic theory of continued fractions and his invention of the (now called) Stieltjes' integral, have had a major impact in all of mathematics. Stieltjes has been rated as one of the chief creators of modern analysis. During his career, he taught at the University of Groningen (The Netherlands) and the University of Toulouse (France). He died in France in 1894. His premature death was a loss for mathematics.

6

Trying Hard to Find Primes

Eric came back, bubbling like a volcano (so it seemed to me), and said:

Eric. You have proven that there are infinitely many primes in the following way, which has merit because of its simplicity.

You said, pick any number $N > 1$, make the number $1 \times 2 \times \cdots \times N$, add 1 to it and you said that any prime p dividing $1 \times \cdots \times N + 1$ must be bigger than N.

What I didn't like is that you did not say if or when $1 \times 2 \times \cdots \times N + 1$ is a prime. So I went home and did some calculations, first adding 1, then subtracting 1, from the product $1 \times 2 \times \cdots \times N$. By the way, before I tell you and discuss my calculations, isn't there a shorter way to write $1 \times 2 \times \cdots \times N$? I'm too lazy to write the product of 100 (or larger number) of factors.

P.P. You are not lazier than anybody else. Mathematicians use the following name and symbol: $1 \times 2 \times 3 \times 4 \times 5 \times 6$ is called the *factorial of* 6 and written 6! In the same way the *factorial of* N is $N! = 1 \times 2 \times \cdots \times N$.

We also like to say, by convention, that the *factorial of* 0 is $0! = 1$.

Eric. I like this notation. So here are parts of my calculations:
$1! + 1 = 2$, prime
$2! + 1 = 3$, prime
$3! + 1 = 7$, prime

$4! + 1 = 25$, not prime, here begins some trouble, but I continued

$5! + 1 = 121$, again not prime.

Will I never find a prime! I kept trying:

$6! + 1 = 721$, not prime

$7! + 1 = 4341$ is?

I tell you, I have no patience to try prime factors less than 70. You still have to teach me something. The way I found out was not very brilliant. Paulo noted that I was in trouble and remembered that he had a copy of that book of records that you spoke badly about. He said he bought it for almost nothing in a bargain basement. The book has a table of primes up to 10 000 and 4341 was not listed. So 4341 is — what do you call a natural number greater than 1, which is not a prime?

P.P. We say it is a *composite* number, because it is the product of smaller factors.

Eric. I'll use that name. On the basis of what I perceived from my calculation, it may well be that only for small numbers N the number $N! + 1$ is a prime. But I don't really know.

P.P. Let us conjugate a verb in the present tense:
    I don't know
    You don't know
    He doesn't know
    She doesn't know
    We don't know
    Nobody knows.

Eric. It is fun how you answered. Can you conjugate it in the future tense?

P.P. Yes, in the future tense, too.

Eric. So you mean you, who are so smart, and all the other brains, cannot say with certainty or even make a guess?

P.P. Prime numbers are hard to study. We have no basis to say anything.

Eric. As I mentioned, I have also calculated $N! - 1$:
$1! - 1 = 0$, not in question
$2! - 1 = 1$, not in question
$3! - 1 = 5$, prime
$4! - 1 = 23$, prime
$5! - 1 = 119$, composite
$6! - 1 = 619$, prime
$7! - 1 = 4339$, prime.

The next number was not in the table, so I stopped. I think it would be very useful to have bigger tables of primes, but it would be even better if you could teach me how to decide if a number is a prime.

P.P. You raise an important question. Again, for $N! - 1$ we have no guess whether we may get infinitely many primes.

Paulo (*who likes big pies and big numbers too*). What are the results of calculations by nuts with supercomputers?

P.P. Before I tell you, I'd like to state, categorically, that the best non-mathematical nuts are the Brazil nuts. They are excellent in pies and other pastries. Let me open that infamous book (that I had to buy at full price!). I learned from you, the next time I want to buy any of MIOBNEBIR's books, I will check bargain basement sales.

Ready for the results? Oh! I cannot find them in the book, which is not very good. However, I can see other tables. Ah! The nuts are not so nuts after all. This is what they do.

There are many variants of the proof of Euclid's theorem. One goes as follows: suppose that $p_1 = 2$, $p_2 = 3$, p_{n+1} is the smallest prime dividing $p_1 \times p_2 \times \cdots \times p_n + 1$. One table contains some values: $p_1 = 2$, $p_2 = 3$, $p_3 = 7$, $p_4 = 43$, $p_5 = 13$, $p_6 = 53$, $p_7 = 5$, $p_8 = 6221671$.

These numbers have been calculated up to p_{43}. But the numbers become too big. Some mathematicians venture that every prime number will be a number in the above sequence of numbers. But this cannot be proven by calculation. If one stops the calculation, say at p_{1000} (which involves huge computations) you would only find 1000 primes. Where to find the others?

There are other calculations of a similar kind. Remember the factorial of N: $N! = 1 \times 2 \times 3 \times \cdots \times N$.

If instead of all natural numbers, I only use primes $p_1 = 2$, $p_2 = 3$, $p_3 = 5, \ldots$. I get these products:
$$p_1\# = 2, \quad p_2\# = 2 \times 3, \quad p_3\# = 2 \times 3 \times 5 \text{ and } p_n\# = p_1 \times p_2 \times \cdots \times p_n.$$

The number $p_n\#$ is called the *primorial* of p_n. There have been calculations of $p_n\# + 1$ and of $p_n\# - 1$. The results are the following:

For primes $p < 120\,000$ all numbers $p_n\# + 1$ have been tested for primality. For the following primes p, $p\# + 1$ is a prime: $p = 2, 3, 5, 7, 11, 31, 379, 1019, 1021, 2057, 3229, 4547, 4787, 11549, 13649, 18523, 23601, 24019$ and 42209.

But the largest known p for which $p\# + 1$ is prime, as known today, is $p = 392113$. The prime number $p\# + 1$ has 169966 digits.

Just try to imagine it!

And do you know what is not known? We don't know if there exist infinitely many primes p such that $p\# + 1$ is a prime. We also don't know if there are infinitely many p such that $p\# + 1$ is composite. But it may well be that there are infinitely many primes such that $p\# + 1$ is a prime, and also infinitely many primes p such that $p\# + 1$ is composite.

Eric. I'm impressed at how easily one encounters such difficult problems.

P.P. I liked when you put the words "easy" and "difficult" side by side. This is common in relation to problems about prime numbers: easy to formulate, but difficult to solve.

Now, Eric and Paulo, do you want to hear about the primality of the numbers $p\# - 1$?

Eric and Paulo (*in total agreement*). Please spare us!

At this point Eric was disappointed, but nevertheless, he persisted in his initial purpose.

Eric. Papa Paulo this is what I want you to teach me. You have to say: "Eric, pick the primes you already have in your pocket, mix them with such and such simple operations so the output is a prime which was not in your pocket. Put this new prime in your pocket and repeat what you did before." Then I will have a pocket full of primes which I know individually.

P.P. I don't know, nobody knows how to indicate simple operators on a finite bunch of primes with an output which is always a prime.

Eric. This is regrettable. I see that primes are wild. If I were a cowboy, I would say that it is not easy to lasso a prime.

Paulo entered in the conversation.

Paulo. I understand that formulas are what we need, formulas which give primes. Can you tell us about formulas for primes?

Eric approved, but I know already that I will disappoint my faithful listeners who asked candid and intelligent questions.

7

A Formula, A Formula, Please

Eric and Paulo asked with a mixture of insistence and desperation: "A formula, a formula, please."

I decided to comply, but had to immediately warn:

P.P. I will discuss formulas for prime numbers, but first I wish to clarify what "formula" means.

Eric. I know this from school. If we want to find some numbers which give the solutions to the problems, we have to replace letters by numbers and do some operations. It is very important to remember the formulas by heart when you go for a test.

P.P. I'll give some formulas, which are not for prime numbers, but they are very interesting and useful.

Paulo. And you'll prove these formulas?

P.P. Yes, I will. The first problem has an interesting, true story. The young Gauss was only 9 years old at the time. One day, the children in the classroom were very noisy. Angrily, the teacher said:

> *Teacher*: You are too noisy! You are not paying attention to the lesson. As punishment, you must add all the numbers $1, 2, 3, 4, \ldots$ up to 100.

And the teacher thought, "Now I'll have some peace and quiet." Not two minutes had passed when young Gauss exclaimed:

> *Gauss*: The answer is 5050.

Amazed with the fast (and correct) answer, the teacher asked Gauss how he found the answer?

Gauss explained: I made believe that I wrote one row with the numbers $1, 2, \ldots$ up to 100, like this:

$$1 \quad 2 \quad 3 \ldots 98 \quad 99 \quad 100$$

Then I imagined a second row below the first one, with the numbers written in the opposite order:

$$1 \quad 2 \quad 3 \ldots 98 \quad 99 \quad 100$$
$$100 \quad 99 \quad 98 \ldots 3 \quad 2 \quad 1$$

Then I added the columns, each time getting 101. There were 100 columns, so the total is $100 \times 101 = 10100$. But each number appeared twice, once in the first row, and again in the second, so I had to divide by 2 and got 5050.

Clever little Gauss. I will tell more about him, as he is considered a great contributor to the knowledge of prime numbers.

The same idea works to find a formula: Let n be any natural number greater than 0, let s_n be a symbol for the sums $1 + 2 + 3 + \cdots + n$. Problem: to find the formula for s_n.

Eric, who was very observant said:

Eric. It is the same as when Gauss had $n = 100$. Do as follows:

$$s_n = 1 + 2 + 3 + \cdots + (n-2) + (n-1) + n$$

and below

$$s_n = n + (n-1) + (n-2) + \cdots + 3 + 2 + 1.$$

Adding by columns

$$2 \times s_n = n \times (n+1)$$

so

$$s_n = \frac{n \times (n+1)}{2}.$$

Paulo. This is a neat formula. So I can see, without big calculations that the sum $1 + 2 + 3 + \cdots + 1000000$ is equal to

$$= \frac{1000000 \times 1000001}{2}$$
$$= 500000 \times 1000001$$
$$= 500000500000.$$

This is really neat. It is good to be smart and tricky like Gauss.

P.P. But you can also be observant and guess the right formula. You just do a few numerical examples until you can make a good guess. Just follow me:

$$s_1 = 1 = 1$$
$$s_2 = 1 + 2 = 3$$
$$s_3 = 1 + 2 + 3 = 6$$
$$s_4 = 1 + 2 + 3 + 4 = 10$$
$$s_5 = 1 + 2 + 3 + 4 + 5 = 15$$
$$s_6 = 1 + 2 + 3 + 4 + 5 + 6 = 21$$
$$s_7 = 1 + 2 + 3 + 4 + 5 + 6 + 7 = 28.$$

By now, we see that if n is even then $n+1$ divides s_n and

$$s_n = (n+1) \times \frac{n}{2}.$$

If n is odd, then n divides s_n and $s_n = n \times \frac{n+1}{2}$.

In both cases, n even or odd, n up to 7, we have $s_n = \frac{n \times (n+1)}{2}$, so we make the guess that the formula is true for an arbitrary n. How about for $n+1$? You write

$$s_{n+1} = s_n + (n+1) = \frac{n(n+1)}{2} + (n+1)$$

hence

$$2 \times s_{n+1} = n \times (n+1) + 2(n+1) = (n+2) \times (n+1)$$

hence

$$s_{n+1} = \frac{(n+1) \times (n+2)}{2}.$$

So the formula that we guessed is also true for the integer $n+1$. We conclude that it is always true.

Eric. Can you find other formulas for sums of natural numbers, when I want to jump numbers?

P.P. I know what you mean. The answer is yes. The problem is the following:

Choose a natural number a, choose $d \geq 1$ and evaluate the sum
$$s_n = a + (a+d) + (a+2d) + (a+3d) + \cdots + (a+(n-1)d)$$
so you jumped $n-1$ times and you have to add up the numbers you touched.

Eric. I am not Gauss but I know what to do.
$$s_n = a + (a+d) + \cdots + a + (n-1)d$$
$$s_n = [a+(n-1)d] + [a+(n-2)a] + \cdots + a$$
$$2s_n = n \times (2a + (n-1)d] = 2an + n(n-1)d$$
hence the formula is
$$s_n = an + \frac{n(n-1)d}{2}.$$
For example, if $a = 3$, $d = 7$ and $n = 5$, we have
$$3 + 10 + 17 + 24 + 31 = 3 \times 5 + \frac{5 \times 4}{2} \times 7 = 85.$$

Paulo. I love this "formula" stuff. Can you show me more formulas?

P.P. As I noted, these formulas have nothing to do with prime numbers. I will show you one more so as not to deviate too much from our principal subject.

This time I will consider the sums of powers of a natural number $a > 1$.

Problem: To find the formula for $A_n = 1 + a + a^2 + a^3 + \cdots + a^n$. It will be an expression involving the numbers a and n. We get the formula with the following trick, or observation: multiply A_n with $a - 1$, we get
$$A_n \times (a-1) = A_n \times a - A_n$$
$$= (a + a^2 + \cdots + a^{n+1}) - (1 + a + \cdots + a^n)$$
$$= a^{n+1} - 1.$$

Therefore $a - 1$ divides $a^{n+1} - 1$ and

$$A_n = \frac{a^{n+1} - 1}{a - 1}.$$

This is the formula. For example, if $a = 3$ and $n = 5$

$$1 + 3 + 3^2 + 3^3 + 3^4 + 3^5 = \frac{3^6 - 1}{3 - 1} = \frac{728}{2} = 364.$$

Paulo. Papa Paulo, you always come up with tricks! How will I ever learn all these tricks?

P.P. As you study, you learn more and more tricks and if you are smart — which you are — you can then invent your own tricks.

Eric. So to solve math problems you must be a tricky magician.

P.P. A little bit of this is true. I can use the formula for A_n and the fundamental theorem to solve a very nice problem. Do you want to hear it?

Paulo. Of course!

P.P. To find a formula for the sum of all factors of a number n. The formula has to be expressed in terms of the factorization of n as a product of primes. I'll derive the formula for an arbitrary natural number $n > 1$ and at the same time work with a numerical example, say, the number 432. We start: n is the product of powers of distinct primes. Let $p_1, \ldots, p_r$ be the distinct prime factors of n; for each prime p_i let $e_i \geq 1$ be such that $p_i^{e_i}$ divides n, but $p_i^{e_i+1}$ does not divide n, so $n = p_1^{e_1} \times \cdots \times p_r^{e_r}$. For $n = 432 = 2^4 \times 3^3$ so $p_1 = 2$, $p_2 = 3$, $e_1 = 4$, $e_2 = 3$. The factors of n are all numbers

$$p_1^{f_1} \times \cdots \times p_r^{f_r} \text{ where } f_1 = 0, 1, \ldots, e_1, \ldots, f_r = 0, 1, \ldots, e_r.$$

In the numerical example, the factors of 432 are all numbers $2^{f_1} \times 3^{f_2}$ where $f_1 = 0, 1, 2, 3, 4$ and $f_2 = 0, 1, 2, 3$, so 432 has $5 \times 4 = 20$ factors. The sum of the factors of 432 is the sum of all numbers $2^{f_1} \times 3^{f_2}$ when f_1, f_2 are as indicated above. We see that $(1 + 2 + 2^2 + 2^3 + 2^4) \times (1 + 3 + 3^2 + 3^3)$ is the sum we want to evaluate — this is seen by multiplication using the distributive property.

But we have seen that
$$1 + 2 + 2^2 + 2^3 + 2^4 = \frac{2^5 - 1}{2 - 1} = 31$$
and
$$1 + 3 + 3^2 + 3^3 = \frac{3^4 - 1}{3 - 1} = \frac{80}{2} = 40.$$

So the sum of the factors of 432 is $31 \times 40 = 1240$. This is the numerical example. To obtain the formula we do the same. The sum of the factions of n is the sum of all numbers $p_1^{f_1} \times \cdots \times p_r^{f_r}$, where $0 \leq f_1 \leq e_1, \ldots, 0 \leq f_r \leq e_r$. The sought-after sum is equal by means of the distributive property of the multiplication, to the product

$$(1 + p_1 + p_1^2 + \cdots + p_1^{e_1}) \times \cdots \times (1 + p_r + p_r^2 + \cdots + p_r^{e_r})$$

which is equal, as we have seen before, to

$$\frac{p_1^{e_1+1} - 1}{p_1 - 1} \times \cdots \times \frac{p_r^{e_r+1} - 1}{p_r - 1}.$$

This is the expression of the sum of factors of $n = p_1^{e_1} \times \cdots \times p_r^{e_r}$.

Paulo. It was a good idea to first explain with numbers.

P.P. If a formula is going to be good for *every* number n, then it has to work for any specific number we choose. I always like to "warm up" by first trying some small numbers.

Now, as I told before, you're going to be unhappy. There are many formulas for prime numbers. You are not the first ones wanting to see such expressions. Some are described, and even proven, in the book by Miobnebir. I wonder why he included these useless formulas. Perhaps just to make his book fat. Since I don't know (and don't want to know) much about them, I peeked in the book. I found not only a formula for primes but also for the number of primes less or equal to n (equally an uninteresting formula).

To indicate the formula for primes, I need to explain the *Möbius function*. This is a function which plays a central role in arithmetic and it is easy to define. Traditionally, the Greek letter μ (read "mu") is used.

$\mu(1) = 1$
$\mu(n) = 1$ if n is the product of an even number of distinct primes
$\mu(n) = -1$ if n is the product of an odd number of distinct primes
$\mu(n) = 0$ if there is a prime p such that p^2 divides n.
I will illustrate giving the smaller values of $\mu(n)$:
$\mu(1) = 1$, $\mu(2) = -1$, $\mu(3) = -1$, $\mu(4) = 0$, $\mu(5) = -1$, $\mu(6) = 1$, $\mu(7) = -1$, $\mu(8) = 0$, $\mu(9) = 0$, $\mu(10) = 1$.
Suppose that $p_1 = 2, p_2 = 3, \ldots, p_{n-1}$ are already known. Let
$$Q = p_1 p_2 \ldots p_{n-1}.$$
This is how one determines the next prime p_n. It is possible to show that there exists a unique natural number $a > 2$ such that
$$2^Q - 1 < 2^{a-1}\big[- (2^Q - 1) + 2\sum\nolimits_d \mu(d)(1 + 2^d + 2^{2d} + \cdots + 2^{(d'-1)d})\big] < 2(2^Q - 1).$$
In the above, $\sum_d$ means a sum where the summands are
$$\mu(d)\big(1 + 2^d + 2^{2d} + \cdots + 2^{(d'-1)d}\big),$$
one summand for each natural number d dividing Q, and $dd' = Q$.

The statement needs of course a proof. Moreover, the number a is just the sought-after prime p_n.

What is your feeling about this formula?

Eric. Bad.

Paulo. Disappointed.

P.P. Worse even. Say you know the first ten primes and you want to determine p_{11} — of course we know that $p_{11} = 31$, but I just want to bring your attention to the kind of calculations that will be needed. The number of summands in the sum $\sum_d$ is equal to $2^{10} = 1024$ and each summand is ...

Eric. The formula is useless for calculation. But is it useful to obtain theorems on prime numbers?

P.P. This has not yet happened, but maybe one day someone will use this idea.

Paulo. Do you think that someday someone, somehow, will discover a simple formula to allow the calculation of primes?

P.P. I don't believe that one such formula can exist.

And the subject of a prime number formula vanished, leaving no hope for a return.

8

Paulo Came with a Lasso

P.P. Paulo, I see that you came today with a lasso. We do not have any cattle here.

Paulo. I am not interested in cows or even buffaloes, but only in primes. I want to lasso some primes.

P.P. If you tell me a natural number, it is always possible in a finite time to tell if the number is a prime, and even to find its factors.

Eric (*intervening*). Big deal! You just try each prime 2, 3, 5, 7 and so on to see if it divides the given number n. As soon as you reach a prime p such that $p^2 > n$, you may stop.

P.P. Good! You remembered what I told you some time ago. In this way, it is guaranteed — I say again — that in a finite time you'll find the factorization of n, including if n is a prime. The big deal, to use your expression, Eric, is the time that will be required. The time depends on how big n is.

The bigger n, the more digits it has, so division requires more operations with the digits. These are called *bit operations*. There are also more primes to test to determine whether they are factors of n. The time will be reduced if you have a fast computer with a large memory, because calculations with very large numbers require increased resources of the computer and need appropriate techniques.

Eric. Suppose I want to find if a number n is prime. If n is so small that it is in tables or published on paper, I can find it right away.

If n is in a table stored in a computer, it's about the same. But if the number n is too big, I would like to know how long it would take to calculate it.

P.P. How many digits do you want in your number?

Eric (*not wanting to seem excessive, said*). 400.

P.P. I wish you a long life Eric, but unless your number is "friendly" or you use a clever method — not trying division by primes — you would be dead before ever knowing the result.

Eric. I plan to live a century, but I still prefer to be lucky and pick a friendly number n and learn what you call a clever method.

P.P. Suppose that someone gives me two primes with about 200 digits each. I hide them from you, multiply these primes, and call the product n; n has about 400 digits. I then ask: Eric, find if n is a prime. I know that it is not, but you don't and you haven't studied clever methods. I also know that n is *not* friendly because n does not have small prime factors. You'll try to divide n by all primes up to the smallest prime p such that $p^2 > n$. So p has roughly 200 digits.

Now I will tell you about an inspired guess by Gauss, of whom I spoke earlier regarding his solution to the sum $(1 + 2 + \cdots + 100)$. According to his guess, which was proven much later, and became a fundamental theorem, the approximate number of primes with at most 200 digits is 6×10^{197}. I am not trying to be accurate, but conveying the number of such primes is written with 197 digits. Just to compare: millions need 9 digits, billions 12 digits, etc.

We have an idea of the number of divisions of the big number n, namely the number of primes we indicated. It is not difficult to calculate from the number of digits how many bit operations will be needed. This is not so important in the present discussion, we just need to be convinced that it is a huge number. Suppose your computer performs 6 000 000 bit operations per second. Much more than 10^{190} seconds will be needed to perform the operations. Not even Superman will live that long. I also have to say that I was very inaccurate in the above calculations. What I want to say is easy to summarize:

(a) Mistrust numbers, some numbers may be very unfriendly.
(b) Learn techniques which allow to treat the problem in a reasonable time.

Paulo. I'll put my lasso aside. You are telling me that I'll only be able to lasso friendly cows or buffaloes.

Before I continued, it occurred to me that Paulo is obsessed with food. In my whole life, I never heard anyone comparing primes to cows!

Eric. I appreciate your calculations of the time necessary to perform a program and solve the problem. But you were obviously not very rigorous.

P.P. There are people who study the number of bit operations, hence the time needed to perform a certain computation. This depends on the number of bits of the given number.

We can say that the time is bigger for bigger numbers, so the time t is a function of the size n. The function time $t(n)$ depends on the algorithm that is used. The best algorithms are those which give a faster answer, so for arbitrary numbers, $n, t(n)$ is smaller. This study is called the *complexity of algorithms*.

Let me explain further how things work. Suppose, Eric, you invent an algorithm to solve an important problem. You are very proud and you want to sell it to a company, so the company will become richer and — of course — reward you with royalties. But you must add to your algorithm a study of its complexity, saying how long it would take to deal with numbers of any given size. You must carefully prove your time estimates.

Paulo. I see that primes have to do with business. I thought they were just numbers to play with, but you can get rich on them.

P.P. This is true and many people, perhaps too many, study primes for lucrative purposes. Why not? If this is what they want, let them do it. As for me, I am interested — that is not the word — I am totally devoted to the study and clarification of the mystery of primes. All this may have applications, even bring me some economic benefits, which I will not refuse, of course. What will your attitude be?

Eric. Definitely to become rich — very rich — after discovering an algorithm with fast performance, then selling it to a large company! Of course, Papa Paulo, I would offer you a generous stipend to continue investigating the mysteries of your favorite numbers.

P.P. I had never thought about that possibility. But it is true that Landon Clay, who always liked mathematics, did not become a professional mathematician, but instead became very wealthy. Like many people, he eventually returned to his first love and through his Clay Institute has been offering US$ 1 000 000 for the solution of each of seven outstanding problems. To tell the truth, the problems are so hard that Clay will not have to write a check during his lifetime.

P.P. Oh look! Time has passed and I did not teach you a thing.

Paulo. You did, I learned about primes, cows, and business.

Eric. And I learned about complexity of algorithms.

P.P. Number theory has been called the *Queen of Mathematics*. Until some fifty years ago, it did not occur to anyone that number theory, especially the study of prime numbers, would have any immediate applications to business. More recently, the Queen has been relegated to be the object of a courtship, inspired by material gains, rather than awe. As a result, progress has been made in unexpected directions, which have required deeper investigations.

Papa Paulo paused and wished to conclude the day's discussion.

P.P. Today I have raised many interesting points — it was a conversation to bring forward difficulties one necessarily finds in studying primes. We shall go back to specific questions and results.

9

Beautiful Old Elementary Arithmetic

P.P. Think about any natural number m greater than 1. I shall say things concerning the remainders of divisions by this number m.

The numbers a and b are called *congruent modulo m* when m divides the difference $a - b$.

Do not forget that m divides 0, so any number a is congruent to itself modulo m.

I show you what happens when m is small. This will help you to understand congruent numbers.

Say $m = 2$. The numbers in the row 0 below are congruent to each other, so are the numbers of the row 1, but the numbers of different rows are not congruent modulo 2.

Row 0: $-6, -4, -2, 0, 2, 4, 6, \ldots$
Row 1: $-5, -3, -1, 1, 3, 5, \ldots$

For $m = 3$, we have three rows like above, in each row the numbers are congruent to each other, but numbers in different rows are not congruent.

Row 0: $-6, -3, 0, 3, 6, 9, \ldots$
Row 1: $-5, -2, 1, 4, 7, 10, \ldots$
Row 2: $-4, -1, 2, 5, 8, \ldots$

You can solve this for $m = 4$, and of course, everything I said can be done for every $m > 1$.

The set of numbers in any one row is called a *residue class* modulo m. The smallest natural number in each residue class is called (not hard to agree) the *least residue* in the class. To be sure that you followed my explanation, I return to the small choices of m. If $m = 2$ there are just two residue classes modulo 2, which are the two rows I had displayed. The least residues are respectively 0 and 1.

For $m = 3$ there are three residue classes, with least residues 0, 1 and 2. You already guessed that for any m there are exactly m residue classes with least residues $0, 1, 2, \ldots, m-1$.

Paulo. And these are exactly the remainders of divisions by m.

P.P. Listen to this: If a and b are natural numbers and if a and b are congruent modulo m, then a and b divided by m have the same remainder. The converse is also true: if a and b divided by m have the same remainder, then a and b are congruent modulo m.

Is this clear, or do you want me to prove it?

Nobody said anything, so I began the proof.

P.P. We divide a by m, getting $a = qm + r$, where the remainder r is such that $0 \leq r < m$. We also divide b by m, getting $b = q_1 m + r_1$, where $0 \leq r_1 < m$. So $a - b = (qm + r) - (q_1 m + r_1) = (q - q_1)m + (r - r_1)$. If a and b are congruent modulo m, then $a - b$ is a multiple of m, but $r - r_1$ lies between $-m$ and m, that is $-m < r - r_1 < m$.

The only multiple of m between $-m$ and m is 0, that is $r = r_1$. The converse is also true, because if the remainders r and r_1 are equal, then $a - b = (q - q_1)m$, so a is congruent to b modulo m.

q.e.d.

You see, Paulo, watch for the remainders.

Gauss — the same Gauss — invented the following notation: $a \equiv b \pmod{m}$, to be read "a is congruent to b modulo m".

It is easy and important to calculate with congruences. What I'm telling now is quite simple:

If $a \equiv b \pmod{m}$ and $c \equiv d \pmod{m}$ then
$$a + c \equiv b + d \pmod{m}.$$

I think that you can prove this one.

Beautiful Old Elementary Arithmetic

If $c = a$ and $d = b$, then we get $2a \equiv 2b \pmod{m}$ and in the same way, $3a \equiv 3b \pmod{m}, \ldots, ka \equiv kb \pmod{m}$ for any number k, also when k is negative.

Concerning multiplication of congruences, we also have:

If $a \equiv b \pmod{m}$ and $c \equiv d \pmod{m}$, then $ac \equiv bd \pmod{m}$.

Eric thought a little and said:

Eric. I can see that $3 \equiv 8 \pmod{5}$ and $7 \equiv 17 \pmod{5}$, then I can see that $3 \times 7 = 21$ is congruent to $8 \times 17 = 136 \pmod{5}$.

How do we do this using letters?

P.P. This proof is also easy. It is clear that if k, h and ℓ are numbers such that $k \equiv h \pmod{m}$ and $h \equiv \ell \pmod{m}$ then $k \equiv \ell \pmod{m}$, because m divides $k - h$ and $h - \ell$, so m divides $(k - h) + (k - \ell) = k - \ell$. I shall use this fact. We have assumed that $a \equiv b \pmod{m}$, so $ac \equiv bc \pmod{m}$. From $c \equiv d \pmod{m}$ we also have $bc \equiv bd \pmod{m}$, this was proven a few minutes ago. Then $ac \equiv bd \pmod{m}$ — this I proved one minute ago. Fast and easy.

Again, taking $c \equiv a$ and $d \equiv b$ we deduce that if $a \equiv b \pmod{m}$ then $a^2 \equiv b^2 \pmod{m}$ and in the same way:

$$a^3 \equiv b^3 \pmod{m}$$

and for each natural number k, $a^k \equiv b^k \pmod{m}$.

We have seen that if $a \equiv b \pmod{m}$ then $ka \equiv kb \pmod{m}$ for any $k \geq 1$. But we have the following: $4 \equiv 14 \pmod{10}$ $4 = 2 \times 2$, $14 = 2 \times 7$, but $2 \not\equiv 7 \pmod{10}$. Of course the symbol $\not\equiv$ means "not congruent".

Nevertheless, we have the very useful fact: if $k > 1$ and m are coprime integers if $ka \equiv kb \pmod{m}$ then $a \equiv b \pmod{m}$.

The proof is again very easy: m divides $ka - kb = k(a - b)$. But m and k are coprime, so, remember, m divides $a - b$, that is, $a \equiv b \pmod{m}$.

Eric wanted to know why congruences were useful.

P.P. Congruences allow us to calculate with remainders, which are usually smaller than the original numbers.

Eric. Give me one example where calculating with the given numbers is time-consuming while calculating with residues is fast. Then I will accept congruences.

P.P. OK, let me think. Here is one: show that $17^{17} + 1$ is divisible by 9. No congruences used: you should multiply 17 by itself 17 times and add 1. Then divide by 9. With congruences: $17 \equiv -1 \pmod{9}$, so $17^2 \equiv (-1)^2 = 1 \pmod 9$, $17^3 \equiv 17 \equiv -1 \pmod 9$ and so on, up to $17^{17} \equiv -1 \pmod 9$, hence $17^{17} + 1 \equiv 0 \pmod 9$, that is, 9 divides $17^{17} + 1$.

Many examples of this kind are possible. And you'll see numerous important theoretical and practical uses of congruences.

Matchbox Cars and Drawers

P.P. Paulo, I know that you have a matchbox car collection.

Paulo. Yes, I do, but we were talking about remainders.

P.P. Here is the point. You are given a carrying case with 35 places and your collection has exactly 35 cars. So you may put one car in each compartment and go around showing the collection. If you have more than 35 cars you are in trouble, you must squeeze at least two cars in one place.

This is also true for pigeons in pigeon-holes and books in drawers. The great mathematician DIRICHLET kept his room in disorder, annoying his — and in fact any — wife: *"Gustav, if you don't clean up your messy room, I'll throw any books lying on the floor straight in the garbage."*

P.P. Between us, how could anyone, even a wife, say such things to a man, whose importance in mathematics will last forever? Dirichlet lived some 150 years ago and is still today admired for his mathematical contributions.

Eric. Most people don't have the faintest idea of what goes on in the brains of mathematicians.

P.P. Indeed, absent-minded, blank stares may suggest a lack of intelligence. It is just the opposite. When I'm thinking hard and look in the mirror I get depressed. But I know that ... well, I confess ... I am smart.

After this reluctant confession, Papa Paulo continued.

P.P. Gustav did not want to lose any of his books. He had 5 drawers and books of 6 different kinds. So, he was forced to have books of two different kinds in the same drawer. This simple event served him well. He invented there and then the famous "drawers principle" which is a key method in some of his proofs. It is called *Dirichlet drawer* (or *pigeon-hole*) *principle*, but could be called the *Paulo matchbox car principle*.

Eric. What is the use of this principle in our discussion?

P.P. I'll state a theorem and even prove it.

Theorem. *Let $m > 1$, let $k \geq 1$ be coprime to m and let r be any number (not excluding 0). We consider the numbers*
$$r, r+k, r+2k, \ldots, r+nk.$$
If $n = m - 1$, the least residues modulo m of these numbers are $0, 1, 2, \ldots, m-1$, written in an appropriate order. If $n \geq m$, then at least two of these numbers have the same least residues modulo m.

Paulo. You don't need to prove it because we already know that if $0 \leq i < j \leq m-1$, then $r + ik \not\equiv r + jk \pmod{m}$, because $0 < j - i < m$ and m, k are coprime. You proved it Papa Paulo, so if $n = m - 1$, we get all possible residues $0, 1, 2, \ldots, m-1$.

The case where $n \geq m$ is just a use of Dirichlet's drawer's principle. I got the idea right away, because of my matchbox cars.

Fermat's Little Theorem

What is next?

P.P. I'll give you a very important theorem discovered by Fermat. First it is instructive to examine simple numerical examples. The theorem is about residue classes modulo a prime number p. We consider any integer $a \neq 0$ such that p does not divide a, then we investigate the residue classes of the powers of a.

We fix $p = 7$, we take $a = 2, 3, 4, 5, 6$ and consider the least residues modulo 7 of their powers

a	a^2	a^3	a^4	a^5	a^6
2	4	1	2	4	1
3	2	6	4	5	1
4	2	1	4	2	1
5	4	6	2	3	1
6	1	6	1	6	1

In this table the least residues are indicated. We observe:

(1) For every a there is a smallest exponent $e > 1$ such that the power a^e has least residue 1 and $e = 2, 3$ or 6.
(2) This smallest exponent e such that $a^e \equiv 1 \pmod{7}$ divides 6 and f is an integer such that $1 \le f$ then $a^f \equiv 1 \pmod{7}$ exactly when e divides f.

You can try the primes $p = 13, 19$ and check that the corresponding facts also hold.

For any odd prime p, we have the so-called:

Fermat's Little Theorem. *Let p be a prime bigger than 2, let a be any integer such that p does not divide a, so $a \not\equiv 0 \pmod{p}$. Then $a^{p-1} \equiv 1 \pmod{p}$.*

I give this proof. Consider the $p - 1$ numbers $a, 2a, 3a, \ldots, (p-1)a$. These numbers are pairwise not congruent modulo p, because p does not divide a. So p does not divide $(j - i)a$ when $1 \le i < j \le p - 1$.

As I said earlier, the least residues of $a, 2a, 3a, \ldots, (p-1)a$ are $1, 2, 3, \ldots, p-1$, written in an appropriate order (of course, I apply the earlier result with $r = 0$ and 0 is the residue of 0). I have also shown that congruences can be multiplied and the result is again a congruence. So

$$a \times 2a \times \cdots \times (p-1)a \equiv 1 \times 2 \times \cdots \times (p-1) \pmod{p}$$

hence
$$a^{p-1} \times 1 \times 2 \times \cdots \times (p-1) \equiv 1 \times 2 \times \cdots \times (p-1) \pmod{p}$$
so
$$(a^{p-1} - 1) \times 1 \times 2 \times 3 \times \cdots \times (p-1) \equiv 0 \pmod{p}.$$

At this point, we observe that p does not divide $1 \times 2 \times \cdots \times (p-1)$. Therefore, from what I have shown earlier, $a^{p-1} \equiv 1 \pmod{p}$.

q.e.d.

Eric. Did you say Fermat's Little Theorem, or Little Fermat's Theorem?

P.P. I said the first. Fermat, unlike Gauss, was not a child prodigy.

Eric. If this is Fermat's Little Theorem there must be a Fermat's Big Theorem.

P.P. Yes, there is such a theorem, which is commonly called Fermat's Last Theorem. That is really interesting. Fermat stated the theorem in 1640, but did not write the proof which, no doubt, he knew, or believed that he knew.

Eric. You made some observations about the residue classes modulo 7. Are such facts true for any prime p?

P.P. Yes, and I plan to explain this to you carefully.

Papa Paulo continued ...

P.P. Our hero today was Fermat.

Paulo. He is not only our hero. I know a very serious looking man who likes Fermat so much that ...

Eric. That the vanity plate of his car is FERMAT.

P.P. Some people want to be famous by any means.

Notes About Pierre de Fermat (1601–1665)

With my friend, I visited Beaumont de Lomagne, a small town situated not far from Toulouse, France. This was the place where Pierre de

Fermat was born. His birth house, now a museum, is of a distinguished architectural style. The market square nearby is still as it used to be, but now has a statue of Fermat. The old church, with its resounding bells, is nearby. From the terrace of Fermat's house one may see the slate roofs of the houses lining the narrow streets.

A good restaurant brought comfort after the enriching visit, with excellent food and wine from the region. This is the place where Fermat grew up and began his studies. Eventually, Fermat became a magistrate in the Parliament of Toulouse. As his main interest, however, he cultivated mathematics. Fermat made discoveries of importance in a variety of fields: geometry of locus of curves extending the work of Appolonius; the theory of tangents of curves, of maxima and minima, which are essential features of differential calculus, but also the calculation of areas, including the basic ideas of integral calculus. With Pascal, Fermat was one of the inventors of probability theory. His work on optics led to a scientific controversy with Descartes. Other work about the free fall of bodies led to the study of spirals and was in certain disagreement with the conclusions of Galileo.

Fermat's special interest was the theory of numbers. Spurred by the reading of Bachet's translation of "*Arithmetica*" by Diophantus of Alexandria, Fermat made a number of interesting discoveries concerning arithmetic. He wrote many notes on his copy of the book. The most famous is the statement which became known as Fermat's Last Theorem: if $n > 2$, there are no integers x, y and z, different from 0, such that $x^r + y^r = z^r$. He added that he had a marvelous proof of this statement, but the margin was too narrow to contain it. Fermat studied the equations of the form $x^2 - Ny^2 = 1$, where N is any positive integer which is not a square. He discovered that the only solution in positive integers of $x^2 + 2 = y^3$ is $x = 5$, $y = 3$. He also discovered that the only solutions in positive integers of $x^2 + 4 = y^3$ are $x = 2$, $y = 2$ and $x = 11$, $y = 5$. Another remarkable discovery was that every prime number congruent to 1, modulo 4, is the sum of two squares. As a rule, Fermat did not write the proofs of his discoveries, which he usually proposed as challenges to contemporary European mathematicians. A noteworthy exception was the proof that the equation $x^4 + y^4 = z^4$ has no solution in non-zero integers.

The story of Fermat's Last Theorem is very interesting. His margin notes became known when, in 1670, Fermat's son Samuel published a new edition of Bachet's books and included Fermat's notes. It is believed that

Fermat himself realized that his proof for arbitrary exponent $n > 2$ was wrong; this explains why he published only the proof when $n = 4$.

The problem exerted a "fatal attraction" on mathematicians. Attempts for its solution led to the creation of new branches in number theory, in particular the theory of cyclotomic fields — and many more. Finally, after more than 350 years, A. Wiles was able in 1994 to complete the proof, a monumental mathematical feat. This problem, outside the topics of our discussions, has been the object of numerous books. The results referred in the discussions about prime numbers include only Fermat's Little Theorem, Fermat's primes and the representations of primes congruent to 1 modulo 4, as sums of two squares.

10

The Old Man Still Knows

The main intention now is to discuss three well-known topics in elementary arithmetic.

P.P. Euler studied Fermat's Little Theorem, realized its importance, and succeeded in proving a theorem which contained Fermat's theorem as a particular case.

When mathematicians get acquainted and study a theorem with substance, it is natural to ask whether the same idea works in a wider context. This is what Euler did and what I will explain to you very carefully.

Eric. Euler — You have never mentioned his name. I know by now that if you mention a name, it must be a mathematician of high level.

P.P. I'll pass you notes on Euler later.

Let me see, Paulo, if you can remember Fermat's little theorem.

Paulo. I liked it so much that I cannot forget it. You start with an odd prime p; it may be any prime, small or large. Then pick any integer a, provided $a \neq 0$ and p does not divide a. Then $a^{p-1} \equiv 1 \pmod{p}$.

P.P. Perfect. Euler thought: instead of a prime p, I'll pick any number $n > 2$, I'll not work with any integer $a \neq 0$, but only with the integers a such that $\gcd(a, n) = 1$. Then I'll look at the powers of a to see what happens.

Soon I'll state and prove Euler's theorem, but I think that it is instructive to precede with some numerical calculations. Eric, pick a number $n \geq 2$.

Eric. $n = 20$.

P.P. I will find the integers a, $1 \leq a < 20$, such that $\gcd(a, 20) = 1$. From $20 = 2^2 \times 5$, it is obvious that the values of a are $a = 1, 3, 7, 9, 11, 13, 17$ and 19. I now list the powers of a, modulo 20, for successive values of $a > 1$:

$a = 3, 9, 7, 1, 3, 9, 7, 1$
$a = 7, 9, 3, 1, 7, 9, 3, 1$
$a = 9, 1, 9, 1, 9, 1, 9, 1$
$a = 11, 1, 11, 1, 11, 1, 11, 1$
$a = 13, 9, 17, 1, 13, 9, 17, 1$
$a = 17, 9, 13, 1, 17, 9, 13, 1$
$a = 19, 1, 19, 1, 19, 1, 19, 1$.

Eric. Very interesting pattern. Can we see what happens with $n = 9$?

P.P. The list of integers a, $1 \leq a < 9$, such that $\gcd(a, 9) = 1$ is 1, 2, 4, 5, 7 and 8. Now I write the powers of each $a > 1$ modulo 9:

$a = 2, 4, 8, 7, 5, 1$
$a = 4, 7, 1, 4, 7, 1$
$a = 5, 7, 8, 4, 2, 1$
$a = 7, 4, 1, 7, 4, 1$
$a = 8, 1, 8, 1, 8, 1$.

For $n = 20$ we have $a^8 \equiv 1 \pmod{20}$; note that 8 is the number of integers a such that $1 \leq a < 20$ and $\gcd(a, 20) = 1$. Actually, for each such a we already have $a^4 \equiv 1 \pmod{20}$.

For $n = 9$, there are 6 integers a such that $1 \leq a < 9$ and $\gcd(a, 9) = 1$. We have $a^6 \equiv 1 \pmod 9$ for each a. We also note that $2^e \not\equiv 1 \pmod 9$ and $5^e \not\equiv 1 \pmod 9$ when $1 \leq e < 6$.

Paulo. I am eager to learn Euler's theorem, the statement and the proof. Will you explain it now?

P.P. Not immediately. First I have to define the integer "phi of n". For the second time, I'll use a Greek letter φ (say "phi"). I do it to conform with the traditional notation. If $n > 1$ then $\varphi(n)$ denotes the number of integers a such that $1 \leq a < n$ and $\gcd(a, n) = 1$. φ is called the *Euler totient function*, or simply the *Euler function*. We have determined that $\varphi(20) = 8$ and $\varphi(9) = 6$.

Eric. I bet you are going to tell us how to compute $\varphi(n)$ for any integer $n > 2$.

P.P. Yes, I will, but Paulo is so keen on listening to Euler's theorem, that I am going to grant his wish.

First Topic: Euler's Theorem

Let $n > 2$, let a be a non-zero integer such that $\gcd(a, n) = 1$. Then $a^{\varphi(n)} \equiv 1 \pmod{n}$.

For simplicity in the following discussion, I will write r instead of $\varphi(n)$. Let $b_1, b_2, \ldots, b_r$ be the r integers such that $1 \leq b_i < n$ and $\gcd(b_i, n) = 1$. We consider the numbers $ab_1, ab_2, \ldots, ab_r$ and we note that these numbers are pairwise incongruent modulo n and also that they are coprime to n. For each $i = 1, \ldots, r$ let c_i be the least residue of ab_i modulo m. Then $c_1, \ldots, c_r$ are pairwise incongruent modulo n and coprime to n. Now remember, the numbers $c_1, \ldots, c_r$ are, apart from a possible rearrangement, the same as $b_1, \ldots, b_r$. So
$$a^r \times b_1 b_2 \ldots b_r = ab_1 \times ab_2 \times \cdots \times ab_r \equiv c_1 c_2 \ldots c_r = b_1 b_2 \ldots b_r \pmod{n}.$$
Hence $(a^r - 1) b_1 b_2 \ldots b_r \equiv 0 \pmod{n}$ so n divides $(a^r - 1) b_1 b_2 \ldots b_r$. But n is coprime to $b_1 b_2 \ldots b_r$, therefore n divides $a^r - 1$. This means that $a^{\varphi(n)} \equiv 1 \pmod{n}$.   q.e.d.

P.P (*to himself*). Bravo, old man, but Bravissimo Euler.

Paulo. Can you compare Euler's and Fermat's theorem?

P.P. Fermat's theorem is about congruences modulo an odd prime p. In this case $1, 2, \ldots, p-1$ are coprime to p, so $\varphi(p) = p - 1$. If you read Euler's statement with $n = p$ it becomes nothing else than Fermat's theorem.

Paulo. So Euler was smarter than Fermat.

P.P. One shouldn't assume that, because Euler came one century later and followed the model of Fermat. What is surprising is that no one in the intervening century found the theorem prior to Euler.

Eric (who does not forgive debts) asked...

Eric. Tell us how to calculate $\varphi(n)$. You already told that $\varphi(p) = p-1$ when p is an odd prime.

P.P. I could also have said that $\varphi(2) = 1$. It is convenient to agree that $\varphi(1) = 1$. Now if I want $\varphi(p^2)$ I observe that with the exception of the $p-1$ numbers $p, 2p, \ldots, (p-1)p$, all numbers a such that $1 \le a < p^2$ are coprime to p^2. So $\varphi(p^2) = (p^2 - 1) - (p - 1) = p^2 - p = p(p-1)$.

Eric. That was easy. How about $\varphi(p^e)$ where $e > 2$?

P.P. I will tell you the result: $\varphi(p^e) = p^{e-1}(p-1)$. You found it easy for $\varphi(p^2)$, so I leave you the task of proving the expression for p^e. To find $\varphi(n)$ for any integer $n > 2$ we proceed as follows:

First find the factorization of n as a product of powers of primes, say it is

$$n = p_1^{e_1} \times p_2^{e_2} \times \cdots \times p_k^{e_k}.$$

Then $\varphi(n) = \varphi(p_1^{e_1}) \times \varphi(p_2^{e_2}) \times \cdots \times \varphi(p_k^{e_k})$ which is equal to

$$\varphi(n) = p_1^{e_1-1}(p_1 - 1) \times p_2^{e_2-1}(p_2 - 1) \times \cdots \times p_k^{e_k-1}(p_k - 1).$$

Try to calculate $\varphi(1800)$.

Eric. Watch what I am doing:

1800	2
900	2
450	2
225	3
75	3
25	5
5	5
1	

so $1800 = 2^3 \times 3^2 \times 5^2$.

By your rule
$$\varphi(1800) = 2^2(2-1) \times 3 \times (3-1) \times 5 \times (5-1) = 480.$$

P.P. You did it right. There are exactly 480 integers a, such that $1 \leq a < 1800$, which are coprime to 1800.

Eric. Your rule is wonderful. But you did not explain how you got it. Can it be proved?

P.P. Of course, otherwise I could not use the rule. I suppose you want to see the proof. It is all a question of counting. Look at my two hands. I can match the fingers of one hand with those of the other hand, so I can say that the two hands have the same number of fingers, in this case 5.

If I have three red pens R_1, R_2, R_3 and two green pens G_1, G_2, I can make exactly 6 pairs of pens of two different colors; note that $6 = 3 \times 2$.

Eric. Is this a proof of anything? I knew that before I was born.

P.P. This is how I will prove the rule for Euler's function:

First Step: I will prove that if h and m are coprime integers greater than 1, then $\varphi(hm) = \varphi(h)\varphi(m)$. I begin as follows:

There are hm least residues r modulo hm; to each r let r_1 be the least residue of r modulo h and let r_2 be the least residue of r modulo m. To r we associate the pair (r_1, r_2). Can another least residue r' modulo hm be also associated to the same pair (r_1, r_2)? This means

$$\begin{cases} r \equiv r_1 \pmod{h} \\ r \equiv r_2 \pmod{m} \end{cases}$$

but also

$$\begin{cases} r' \equiv r_1 \pmod{h} \\ r' \equiv r_2 \pmod{m} \end{cases}.$$

So $$r \equiv r' \pmod{h}$$
and $$r \equiv r' \pmod{m}.$$

Therefore both h and m divide $r - r'$. But h and m are coprime, so hm divides $r - r'$, that is $r \equiv r' \pmod{hm}$. Now, remember r and r' are least residues modulo hm, therefore $r = r'$.

Next I note that there are hm residue classes modulo hm and just the same number of pairs of residue classes modulo h and modulo m.

Good, now I take only the $\varphi(hm)$ least residues r which are coprime to hm. Then

$$\gcd(r_1, h) = 1 \text{ and } \gcd(r_2, m) = 1;$$

vice-versa if r is such that $\gcd(r_1, h) = 1$ and $\gcd(r_2, m) = 1$, then r is coprime to hm. This tells me that the matching implies that $\varphi(hm) = \varphi(h)\varphi(m)$. So you see, in the first step I needed to count.

Second Step: Let $n = p_1^{e_1} p_2^{e_2} \ldots p_k^{e_k}$ with $k \geq 2$, $h = p_1^{e_1}$. $m = p_2^{e_2} \ldots p_k^{e_k}$, so $n = p_1^{e_1} m$ where $m = p_2^{e_2} \ldots p_k^{e_k}$, so $\gcd(p_1^{e_1}, m) = 1$. Then $\varphi(n) = \varphi(p_1^{e_1})\varphi(m)$. Repeating the argument for m, we get the proof of our rule. q.e.d.

Paulo. Papa Paulo, you have the gift of making proofs so natural, they do exactly what they are expected to do. Counting, repeating arguments. Great!

Second Topic: The Pairing of Residue Classes

P.P. What I'll say in the next theorem is that certain least residues a may be paired with a least residue a' — and only one — as if a' would be the best friend of a.

Eric. So Papa Paulo, you are really going to prove a theorem about least residues. What is the theorem?

P.P. I will state and prove the theorem.

Theorem. *Let n be greater than 2 and let a be such that $1 \leq a < n$ and $\gcd(a, n) = 1$. Then there exists an integer a', $1 \leq a' < n$ such that $aa' \equiv 1 \pmod{n}$. Moreover, a' is the only integer with this property.*

Proof: It's as easy as it can be. Since $\gcd(a,n) = 1$, there exist non-zero integers s and t such that $as + nt = 1$. Let a' be the least residue of s modulo n, so $s \equiv a' \pmod{n}$. Therefore $aa' \equiv as \equiv 1 \pmod{n}$.

The second part is even easier to prove. If a'' is also an integer such that $1 \leq a'' < n$ and $aa'' \equiv 1 \pmod{n}$, I will show that $a'' = a'$. OK! n divides $aa' - 1$ and n divides $aa'' - 1$, so n divides the difference $(aa' - 1) - (aa'' - 1) = aa' - aa'' = a(a' - a'')$. But n is coprime to a, so n divides $a' - a''$. Since $1 \leq a' < n$ and $1 < a'' < n$ then for sure $-n < a' - a'' < n$. It follows that $a' - a''$ has to be 0, that is $a' = a''$. q.e.d.

What happens when $n = p$ is an odd prime?

Paulo. You just read the theorem. If $1 \leq a < p$ there exists a unique integer a' such that $1 \leq a' < p$ and $aa' \equiv 1 \pmod{p}$. I can also show how to find a', when a and n are given. In fact, we did it already and it is indicated in the proof. Papa Paulo, I take $n = 1000$ and $a = 7$. I calculate the greatest common divisor of 7 and 1000. If it turned out not to be 1, I would not know how to continue. But it is 1 and we learned how to find s, t such that $7s + 1000t = 1$.

Give me one minute and I will calculate the result.

Third and Last Topic for Today: The Theorem of Wilson

As soon as the name Wilson was pronounced, Eric said:

Eric. President Wilson? Of the United States? I did not know that he had made a discovery in mathematics.

P.P. My Wilson is not the President Wilson of the United States, mine was John Wilson, an English mathematician who lived in the 17th century. His theorem is about very large numbers called *factorials*. But Wilson made no calculations in his discovery — it was sheer intelligence that allowed him to prove the theorem.

After this aside, Papa Paulo said:

P.P. I will divide the theorem of Wilson into two parts. In the first part, I will tell you what happens for prime numbers. In the second part, I will deal with composite integers. We shall use $(p-1)!$ to mean the product of all integers from 1 to $(p-1)$. And this is usually called the *factorial* of $(p-1)$.

Theorem. (1) *If p is a prime number, then $(p-1)! \equiv -1 \pmod{p}$.*

(2) *Let n be a natural number greater than 1. If n is composite, then $(n-1)! \not\equiv -1 \pmod{n}$.*

Proof of part (1): In the proof, I shall use facts that I proved earlier. We have $(2-1)! \equiv -1 \pmod{2}$ and $(3-1)! \equiv -1 \pmod{3}$ so part (1) is true for $p = 2$ and $p = 3$. Now we assume that $p \geq 5$ and we consider the $p-3$ numbers $2, 3, \ldots, p-2$. If $2 \leq a \leq p-2$ we have seen that there exists a', unique such that $1 \leq a' \leq p-1$ and $aa' \equiv 1 \pmod{p}$; we see that $a' \neq 1$ and $a' \neq p-1$. Thus the $p-3$ numbers $2, 3, \ldots, p-2$ are grouped in pairs (a, a'), where $aa' \equiv 1 \pmod{p}$. Thus $2 \times 3 \times \cdots \times (p-2) \equiv 1 \pmod{p}$ and $(p-1)! = 1 \times 2 \times \cdots \times (p-2) \times (p-1) \equiv p-1 \equiv -1 \pmod{p}$.

Proof of part (2): Let m be such that $1 < m < n$ and m divides n; then m divides $(n-1)!$, so $(n-1)! \equiv 0 \pmod{m}$.

If we also had $(n-1)! \equiv -1 \pmod{n}$, then n divides $(n-1)!+1$, so m divides $(n-1)!+1$, hence $(n-1)! \equiv -1 \pmod{m}$ and this is a contradiction, concluding the proof. q.e.d.

Eric listened to the proof and made a pertinent observation:

Eric. Papa Paulo, give me a number $n > 1$. I calculate $(n-1)!$ modulo n. If I get -1 then n is a prime, if I don't get -1 then n is composite. What is the fuss about deciding if a number is a prime?

P.P. Up to the moment you asked "what is the fuss ...," I was in agreement with you. Let us try one number — not even a big number — say 101.

Eric. I'll do my calculations in front of you.

I watched Eric's diligent work, which he disposed very neatly. He had to determine the least residue of 100! modulo 101:

    1
    2
    6
    24
    $5 \times 24 = 120 \equiv 19 \pmod{101}$
    $6 \times 19 = 114 \equiv 13 \pmod{101}$
    $7 \times 13 = 91$
    $8 \times 91 = 728 \equiv 21 \pmod{101}$
    $9 \times 21 = 189 \equiv 88 \pmod{101}$
    $10 \times 88 = 880 \equiv 72 \pmod{101}$.

As the calculations continued, I noticed that Eric was becoming less enthusiastic. Then he said...

Eric. Now I understand the fuss. Too many calculations, even for 101, which is a small number.

Paulo. Once again, I loved the theorem and the clever proof.

Notes About Leonhard Euler (1707–1783)

In 2007 the scientific world commemorated the 300th anniversary of the birth of Euler, one of the greatest mathematicians of all time. Born and educated in Basel, Switzerland, Euler studied philosophy, theology, Greek, Hebrew, and mathematics at the University of Basel, where he also attended the lectures of Johann Bernoulli, of the famous scientific family. At age 16, Euler defended his Master's thesis on philosophy concerning Descartes, Newton, and Galileo. At age 19 he finished his mathematics studies at the university and began to publish research papers. An essay about naval engineering won the second prize in a contest promoted by the Académie des Sciences de Paris. At age 20, Euler wrote a paper on acoustics, which became a classic. Euler's friend Daniel Bernoulli offered him a position in St. Petersburg and by 1730, Euler had become a professor of physics at the Academy of St. Petersburg. Upon the departure of Daniel Bernoulli in 1733, Euler was appointed professor of mathematics.

Euler's rise was meteoric and he soon acquired a great reputation among mathematicians and scientists in general. He stayed in St. Petersburg until 1744, when he was appointed by King Frederick the Great of Prussia to create and direct the Berlin Academy of Sciences. In 1766 he was recalled to St. Petersburg by Catherine the Great, to become the director of the Academia Petropolitana, a position which he occupied until his death in 1783.

Sometimes marriage and children help mathematics. Euler had one wife and 13 children, of which only (!) 5 survived infancy. Euler once said that he made some of his most important mathematical discoveries holding a baby in his arms, with children playing at his feet.

In 1738, Euler became blind in his right eye, and in 1767, cataracts made him blind in his left, but this did not seem to be a great obstacle to him. Despite this severe handicap, Euler continued his research work and performed his duties in the Academy. In particular, Euler dictated the book "*Algebra*" to his valet at the end of his life.

Euler was a prolific author. At present, his published collected papers occupy 39 substantial volumes. There is a large body of work devoted to analysis, to geometry, to the study of series, to arithmetic, applied mathematics, acoustics, optics, but also papers on cartography, and naval engineering.

In our discussions we shall only encounter a limited number of Euler's discoveries: the number e, with $\log e = 1$, the extension of Fermat's little theorem, the function $\varphi(n)$, various properties of residue classes, the zeta function and a variety of other results on divisibility.

11

Can You Tell Me All About Congruences?

P.P. We need more facts about congruences. Due to their importance, I will also mention some results which will not be needed. But do not forget: I am ready to answer any questions.

Eric. Remember when we calculated powers of residues modulo 20 and modulo 9? We made some interesting observations; you said that you would treat the powers of residues of any modulus.

P.P. This I can do for you. Paulo, what do you want to hear?

Paulo. I am curious — but I don't know why — about residue classes which are squares. When I was playing with the residue classes modulo 7, I found that $2 \equiv 3^2 \pmod 7$, so the least residue 2 is a square modulo 7. But the number 2 is not a square. I would like you, Papa Paulo, to explain all about square congruence classes.

P.P. This is a very interesting phenomenon, which I will discuss in detail.

Making Eric Happy

P.P. First I'll take care of Eric's request. But, Paulo, you also have to hear what I will say.

P.P. If $n > 2$ and $\gcd(a, n) = 1$, by Euler's theorem $a^{\varphi(n)} \equiv 1 \pmod n$. So there exists the smallest integer $e \geq 1$ such that $a^e \equiv 1$

(mod n). Thus $e \leq \varphi(n)$. This integer e is called the *order of a modulo* n. But there are also other integers f such that $e \leq f$ and $a^f \equiv 1 \pmod{n}$. These numbers f are easy to determine:

Theorem. *With a, n and e as above, $a^f \equiv 1 \pmod{n}$ only if e divides f.*

Proof: We have to show that if $f = qe + r$ where $0 \leq r < e$, then $r = 0$. By assumption, $1 \equiv a^f \equiv a^{eq+r} \equiv (a^e)^q \times a^r \equiv 1 \times a^r \equiv a^r$ (mod n). Remember $0 \leq r < e$ and e is the smallest integer $e \geq 1$ such that $a^e \equiv 1 \pmod{n}$. So, r must be 0. q.e.d.

This is what happened with $n = 20$ and $n = 9$. From our calculations 1 has order 1 modulo 20; $3, 7, 9, 11, 17$ have order 4 modulo 20; 19 has order 2 modulo 20. No a has order $\varphi(20) = 8$.

For the modulus 9, 1 has order 1 modulo 9; 2 and 5 have order $6 = \varphi(9)$; 4 and 7 have order 3 modulo 9; 8 has order 2 modulo 9.

Now I want to mention an important concept:

The integer g such that $1 < g < n$ and $\gcd(g, n) = 1$ is called a *primitive root modulo* n if the order of g modulo n is $\varphi(n)$, which is the maximum possible.

Eric (*complaining*). Aye, aye, aye! If $n = 20$, there is no primitive root modulo 20. So there is a flaw in your definition. You are talking about nothing!

P.P. Excellent remark. But there is no problem. Here is how it goes:

We defined the primitive root for any modulus. We know that if $n = 20$ there is no primitive root modulo 20, but there are primitive roots modulo 9. So it is important to know for which moduli n there is a primitive root modulo n.

I can state — but will not prove — this theorem:

Theorem. *There is a primitive root modulo 2 and 4, and when $n = p^k$ or $2p^k$, where p is an odd prime and $k \geq 1$.*

Eric. I want to believe you. But you should be more convincing! Tell me: If I pick a prime, is there a method to find a primitive root modulo p?

P.P. Gauss (*him again!*) invented a method with which mathematicians could prove how to estimate the expected size of the smallest primitive root g. I am not going to relate it. You may find the result in books and papers; I checked Miobnebir's book to see if he talks about it, and he does. But you know what I would do? Just try $2, 3, \ldots$ small numbers to see if you are able or lucky enough to find a primitive root. Miobnebir's book has a short table of primitive roots; I will give some values for prime moduli p up to 100. 2 is the smallest primitive root modulo $p = 3, 5, 11, 13, 19, 29, 37, 53, 59, 61, 67$ and 83. Again, 3 is the smallest primitive root modulo $p = 7, 17, 31, 43, 79$ and 89. Try 3 modulo 31.

Eric. Nothing to it. I just write the sequence of powers of 3 modulo 31. You know, I may replace $27 = 3^3$ by -4 because $27 \equiv -4 \pmod{31}$. Here it is:

$$3, 9, \ 27 \equiv -4, -12, \ -36 \equiv -5, -15, \ -45 \equiv -14,$$
$$-42 \equiv -1, \ -33 \equiv -2, -6, \ 42 \equiv 11, \ -18 \equiv 13,$$
$$39 \equiv 8, \ 24 \equiv -7, \ -21 \equiv 10, \ 30 \equiv -1.$$

Now I stop, because I got $3^{15} \equiv -1 \pmod{31}$; I also know that $3^{30} \equiv 1 \pmod{31}$. By Fermat's theorem, the order of 3 modulo 31 has to divide 30, so it must be 30, that is, 3 is a primitive root modulo 31.

Eric again:

Eric. I like to do as little work as possible! I laugh when g is small, say $g = 2$. I noted that, among the 24 odd primes $p < 100$, 2 is a primitive root for 12 of these primes — half of the possible 24 primes. I feel tempted to guess that there are infinitely many primes p such that 2 is a primitive root modulo p. Is this true?

P.P. Nobody knows the answer to that question. One day, I shall discuss mysteries of prime numbers and consider the problem.

Eric. Why do you like primitive roots so much?

P.P. I will explain with the modulus 11. First I write the powers of 2, then of 3 modulo 11 up to the 10th power

$2, 4, 8, 16 \equiv 5, \quad 10, 20 \equiv 9,$
$18 \equiv 7, \quad 14 \equiv 3, \quad 6, 12 \equiv 1.$

$3, 9, 27 \equiv 5, \quad 15 \equiv 4, \quad 12 \equiv 1,$
$3, 9, 27 \equiv 5, \quad 15 \equiv 4, \quad 12 \equiv 1.$

What do you see? 2 is a primitive root modulo 11, the least residues of the powers of 2 are, in an appropriate order, the 10 integers 1, 2, 3, 4, 5, 6, 7, 8, 9, 10.

On the other hand, 3 has order 5, so it is not a primitive root modulo 11, the least residues of the powers of 3 are just 1, 3, 4, 5 and 9 in an appropriate order.

The same is true for any modulus p and primitive root g.

If p is an odd prime, if g is a primitive root modulo p, the least residues of the powers $g, g^2, g^2, g^3, \ldots, g^{p-1}$ are, in an appropriate order, $1, 2, \ldots, p-1$.

Papa Paulo paused and spoke again:

Patient Paulo's Turn

P.P. Paulo, thanks for your patience. It is your turn and I will address your question.

To warm up, let's start with something easy. Question: Which are the integers a such that $a^2 = 1$.

Paulo. Just $a = 1$.

P.P. Not just ... you forgot that according to the rule of signs $(-1) \times (-1) = 1$, so $(-1)^2 = 1$. Now we consider the same question modulo an odd prime p. In this case $-1 \equiv p - 1 \pmod{p}$. The result is...

Paulo interrupted me and ventured...

Theorem. *Let p be an odd coprime, let a be an integer such that $1 \leq a \leq p - 1$. If $a^2 \equiv 1 \pmod{p}$ then $a = 1$ or $p - 1$.*

P.P. That is the theorem. Can you prove it?

Paulo. Let me try. I have

$$a^2 \equiv 1 \pmod{p},$$

so
$$0 \equiv a^2 - 1 = (a+1)(a-1) \pmod{p}.$$

Thus p divides $(a+1)(a-1)$, therefore p divides $a+1$ or p divides $a-1$. In the first case, $a = p-1$ and in the second case, $a = 1$.

q.e.d.

Paulo. This was easy. There is no way you can make a mistake. But his little theorem (why don't you call it Little Paulo's Theorem?) is not an answer to my question. Tell me when an integer is a square modulo p. Forget about 0, 1, which are clearly squares. How about other integers?

P.P. To understand the squares modulo p, we choose any primitive root g modulo p. We already know two things: $g, g^2, \ldots, g^{p-1}$ have different least residues modulo p, and if $1 \le k < h$ and $g^k \equiv g^h \pmod{p}$ then $h \equiv k \pmod{p-1}$.

Paulo complained:

Paulo. You never gave us the second one. But I think I can figure it out. I write $h = q(p-1) + r$ with $0 \le r < p-1$, I also write $k = q_1(p-1) + r_1$ with $0 \le r_1 < p-1$. Then $g^h \equiv (g^{p-1})^q \times g^r \equiv 1 \times g^r \equiv g^r \pmod{p}$ and also $g^k \equiv (g^{p-1})^{q_1} \times g^{r_1} \equiv 1 \times g^{r_1} \equiv g^{r_1} \pmod{p}$. From $g^k \equiv g^h \pmod{p}$ I deduce that $g^r \equiv g^{r_1} \pmod{p}$, where $0 \le r < p-1$ and $0 \le r_1 < p-1$. If $r = 0$ then $1 \equiv g^r \equiv g^{r_1} \pmod{p}$, so this forces $r_1 = 0$. If $r \ne 0$ then also $r_1 \ne 0$, and because g is a primitive root modulo p, then $r = r_1$ also in this case. Thus $k \equiv h \pmod{p-1}$.

P.P. Let me prove an easy theorem. By the way, theorems which are used to prove important theorems are normally called *lemmas*. So I'll use this name:

Lemma. *Let g be a primitive root modulo p. The least residue of the power g^k (where $0 \le k$) is congruent to a square modulo p if, and only if, k is even.*

Paulo. This will make it easy to find the squares.

Eric added:

Eric. Provided that you know how to determine a primitive root modulo p. Papa Paulo, you have not explained this to us. I understand why Gauss was so keen on having a method to find primitive roots.

Paulo. Papa Paulo, I hope you are not skipping the proof of the lemma.

P.P. Not skipping. Here is the proof. I shall use the notation, for any $k \geq 0$: a_k denotes the least residue of g^k modulo p, so $g^k \equiv a_k$ (mod p). Hence $g^{2k} \equiv a_k^2$ (mod p). This was the proof of one assertion of the lemma.

For the converse, let $h \geq 0$, and assume that there is a least residue a_k modulo p, such that $g^h \equiv a_k^2$ (mod p). From $a_k^2 \equiv g^{2k}$ (mod p) then $g^h \equiv g^{2k}$ (mod p), therefore $h \equiv 2k$ (mod $p-1$). That is, $h = 2k +$ multiple of $p-1$. Hence h is even. q.e.d.

Paulo. Easy! Now, tell us more about squares modulo p, in particular when -1 is a square modulo p. I find this unreal — no, let me take that back — I would have found it strange that -1 is a square, because the square of any non-zero integer is positive. But of course, we are now considering residue classes modulo p. And I noticed that if $p = 17$ then
$$-1 \equiv 16 \equiv 4^2 \pmod{17}.$$
Such relations must happen more times. I am curious about when -1 is a square modulo p.

P.P. Let me first do something which will be useful.

Theorem. *Let p be an odd prime, let a be an integer not a multiple of p. Then a is a square modulo p if and only if $a^{\frac{n-1}{2}} \equiv 1$ (mod p).*

Proof: First we assume that a is a square modulo p, so there exists an integer b such that $a \equiv b^2$ (mod p). It follows that
$$a^{\frac{p-1}{2}} \equiv b^{2\times\frac{n-1}{2}} \equiv b^{p-1} \equiv 1 \pmod{p}$$
as said by Fermat's Little Theorem.

Now let $a^{\frac{n-1}{2}} \equiv 1 \pmod{p}$. Let g be a primitive root modulo p, so there exists an integer h such that $1 \leq h \leq p-1$ and $a \equiv g^h \pmod{p}$.

Therefore, $1 \equiv a^{\frac{n-1}{2}} \equiv g^{h \times \frac{n-1}{2}} \pmod{p}$. Since g is a primitive root modulo p, then $p-1$ divides $h \times \frac{n-1}{2}$. As a consequence h is even. Then $a \equiv g^h \pmod{p}$ implies that a is a square modulo p.

q.e.d.

P.P. The assertion is an easy consequence of the lemma. As such, it is called a *corollary*.

Paulo. What do you say when $a \equiv -1 \pmod{p}$?

Corollary. *Let p be an odd prime. Then -1 is a square modulo p if, and only if, $p \equiv 1 \pmod{4}$.*

Paulo. Don't prove the corollary. I'll do it. If -1 is a square modulo p, then $(-1)^{\frac{n-1}{2}} \equiv 1 \pmod{p}$. But $(-1)^{\frac{n-1}{2}}$ is equal to 1 when $\frac{p-1}{2}$ is even, and equal to -1 when $\frac{n-1}{2}$ is odd. Odd is impossible, it would give $-1 \equiv 1 \pmod{p}$. So, $\frac{p-1}{2}$ is even, hence $p \equiv 1 \pmod{4}$.

The converse is just as easy with what we know. We have $\frac{p-1}{2}$ is even, say equal to $2h$. Then $(-1)^{\frac{p-1}{2}} = (-1)^{2h} = 1$, therefore -1 is a square modulo p, as it was shown in the theorem. q.e.d.

Eric. In my head I was thinking: We need primes $p \equiv 1 \pmod{4}$. We begin with 1, jump by 4 and see if we get primes:

1 5 9 13 17 21 25 29 33 37 41 45 49 53 57 61 65.

It is a little like the Eratosthenes Sieve — not quite the same. Several numbers in the sequence are primes, others are not. Still, I bet — correct me if I am wrong — that in the sequence there are infinitely many primes. This amounts to saying that -1 is a square modulo infinitely many primes.

Eric stopped, and I said:

P.P. At this point I will say only that you are right, but I promise that I'll come back to this question one day.

Paulo. Tell us when 2 is a square modulo p.

P.P. This is the theorem, which I will not prove:

Theorem. *2 is a square modulo the odd prime p, if and only if $p \equiv 1$ or $-1 \pmod{8}$.*

I am sorry. I have to refrain from proving everything. But I can illustrate with some choices of p. In each case, I choose a primitive root g modulo p, and determine h such that $2 \equiv g^h \pmod{p}$. According to h being even or odd, I deduce that 2 is a square or is not a square modulo p

$p = 3$, $g = 2$, so $h = 1$ and 2 is not a square modulo 3;
$p = 11$, $g = 2$, so $h = 1$ and 2 is not a square modulo 11;
$p = 19$, $g = 2$, $h = 1$ and 2 is not a square modulo 19;
$p = 43$, $g = 3$, the powers of 3 modulo 43 are:
3, 9, 27, −5, −15, −2, −6, −18, −11, 10, 30, 4, 12, −7, −21, −20, −17, −8, −24, 14, −1, −3, −9, 16, 5, 15, $2 \equiv 3^{27} \pmod{43}$.

So again, 2 is not a square modulo 43. We observe that 3, 11, 19 and 43 are in the same residue class modulo 8. If you like, you may try $p = 7, 23, 31$ and 47. According to the theorem, 2 will be a square modulo for these primes.

Paulo. If I start with 1 (or 3, or 5, or 7) and jump by 8, I will hit as many primes as I want — not infinitely many primes as I would be dead long before — provided that I keep jumping by 8 and my legs are not too tired.

P.P. This is true, and an instance of what Eric indicated.

Paulo. Now we know the primes p for which -1 and 2 are squares modulo p. What can you say for an arbitrary integer (including negative integers) a, not a multiple of p?

P.P. Many things are easy. To begin, let p be an odd prime and g a primitive root modulo p. Let a be an integer, not a multiple of p. If a is a square modulo p we say that a is a *quadratic residue* modulo p. Following Legendre, it is customary to write $\left(\frac{a}{p}\right) = 1$. If a is not a square modulo p, we say that a is a *non-quadratic residue* modulo p and we write $\left(\frac{a}{p}\right) = -1$. The symbol $\left(\frac{a}{p}\right)$ is called the *Legendre* *symbol* and its values are 1 or -1. Do you agree with what I am saying?

$\left(\frac{a}{p}\right) = 1$ if and only if $a \equiv g^{2k} \pmod{p}$, for some integer k and $\left(\frac{a}{p}\right) = -1$ if and only if $a \equiv g^{2k+1} \pmod{p}$ for some integer k.

Paulo nodded "yes".

P.P. If a is a square then $\left(\frac{a}{p}\right) = 1$.

Paulo nodded "yes" again.

P.P. If $a \equiv b \pmod{p}$ then $\left(\frac{a}{p}\right) = \left(\frac{b}{p}\right)$, so if a is a natural number, writing $a = qp + r$ with $0 \leq r < p$, then

$$\left(\frac{a}{p}\right) = \left(\frac{r}{p}\right).$$

Paulo nodded "yes" once more.

P.P. If a is a natural number which is not a square, then a is the product of a square and a product of distinct primes.

As Paulo did not nod "yes" immediately I added, as an example

$$a = 2^5 \times 3^4 \times 5 \times 7^3 = (2^2 \times 3^2 \times 7)^2 \times 2 \times 5 \times 7.$$

Now Paulo nodded "yes".

P.P. Putting together what I said:

If one knows $\left(\frac{-1}{p}\right)$ (and we do), $\left(\frac{2}{p}\right)$ (we do also) and $\left(\frac{q}{p}\right)$ for all odd primes $q < p$, then we may calculate $\left(\frac{a}{p}\right)$ for all integers a, not divisible by p.

Paulo had to nod "yes", but asked:

Paulo. How does one calculate $\left(\frac{q}{p}\right)$?

P.P. Gauss solved this problem while still a teenager. It was the most striking discovery in the theory of congruences and is expressed in the *quadratic reciprocity law*. Gauss himself gave six different proofs of this law. Many, many authors gave subsequent proofs; the literature counts more than 160 of them.

Paulo. And I gather that you are not telling us even one of these proofs. Why?

P.P. Each one of the proofs requires a certain amount of explanations which are more appropriate in university courses than in our friendly discussions. But I'll state and explain the law, and use it in numerical examples.

The purpose of the quadratic reciprocity laws is to find out if $\left(\frac{q}{p}\right)$ is equal to 1 or to -1 — here $2 < q < p$ and q and p are primes. Why is the law a "reciprocity law"?

There was silence, so Papa Paulo continued.

P.P. The value of $\left(\frac{q}{p}\right)$ depends on the value of $\left(\frac{p}{q}\right)$.

Quadratic reciprocity law: If $2 < q < p$, where q and p are primes, then

$$\left(\frac{q}{p}\right) = \left(\frac{p}{q}\right) \text{ when } p \equiv 1 (\text{mod } 4)$$

or $q \equiv 1 (\text{mod } 4)$

and

$$\left(\frac{q}{p}\right) = -\left(\frac{p}{q}\right) \text{ when } p \equiv 3 (\text{mod } 4)$$

and $q \equiv 3 (\text{mod } 4)$.

Now I will work out a numerical example to find out if $-2^5 \times 3^4 \times 5 \times 7^3$ is a square modulo $p = 1319$. The calculation is:

$$\left(\frac{-2^5 \times 3^4 \times 5 \times 7^3}{1319}\right) = \left(\frac{-2 \times 5 \times 7}{1319}\right)$$

$$= \left(\frac{-1}{1319}\right) \times \left(\frac{2}{1319}\right) \times \left(\frac{5}{1319}\right) \times \left(\frac{7}{1319}\right).$$

We calculate each one of the four Legendre symbols:

$$\left(\frac{-1}{1319}\right) = -1 \quad \text{because} \quad 1319 \equiv 3 \pmod 4.$$

$$\left(\frac{2}{1319}\right) = 1 \quad \text{because} \quad 1319 \equiv 7 \pmod 8.$$

$$\left(\frac{5}{1319}\right) = \left(\frac{1319}{5}\right) = \left(\frac{4}{5}\right) = 1.$$

$$\left(\frac{7}{1319}\right) = -\left(\frac{1319}{7}\right) = -\left(\frac{3}{7}\right) = -\left(-\left(\frac{7}{3}\right)\right) = \left(\frac{7}{3}\right)$$

$$= \left(\frac{1}{3}\right) = 1.$$

Putting together

$$\left(\frac{-2^5 \times 3^4 \times 5 \times 7^3}{1319}\right) = (-1) \times 1 \times 1 \times 1 = -1,$$

so the given number is not a quadratic residue modulo 1319.

Paulo. How fascinating! Can I practice some examples?

P.P. Eric, I want you to do these too. Compute the three Legendre symbols $\left(\frac{25432}{8227}\right)$, $\left(\frac{-80022}{9973}\right)$, $\left(\frac{4577}{5309}\right)$. While you are doing these computations, you will have to find factors of numbers less than the given primes p. These primes p have four digits. All you need to do is to try primes q up to 100 and check if any one is a factor. You can also use a trick — which I will teach you — to reduce size. If a is odd, then $3 \times a, 5 \times a, \ldots$ are odd. If $\frac{p}{3} < a < p$ then the integer $b = \frac{p-a}{2}$ is such that $b < a$ and we have $\left(\frac{a}{p}\right) = \left(\frac{-2b}{p}\right) = \left(\frac{-1}{p}\right)\left(\frac{2}{p}\right)\left(\frac{b}{p}\right)$. If $\frac{p}{5} < a < \frac{p}{3}$ then $b = \frac{p-3a}{2} < a$ and again

$$\left(\frac{a}{p}\right) = \left(\frac{3}{p}\right)\left(\frac{3a}{p}\right) = \left(\frac{3}{p}\right)\left(\frac{-1}{p}\right)\left(\frac{p-3a}{p}\right)$$

$$= \left(\frac{3}{p}\right)\left(\frac{-1}{p}\right)\left(\frac{2}{p}\right)\left(\frac{b}{p}\right).$$

You may figure out the trick for other values of a.

Something may happen in your calculations: If you make *one* mistake the result is 1 when it should be -1, or -1 when it should be 1. So it is wrong. What happens if you make *two* mistakes?

Paulo. It is twice wrong.

P.P. Yet it gives the right result!

Paulo. This is really fun.

P.P. My problems will require calm and concentration to avoid errors.

Eric. Tell me how to check if I did not make any errors.

P.P. You could have made *one* error and, in repeating the errors, *three* errors getting in both cases a wrong result. To be sure, it is necessary to use a completely different method of checking. First you determine a primitive root g modulo p. Then you find k, $1 \le k \le p-1$, such that $a \equiv g^k \pmod{p}$. If $k = 2h$ then $\left(\frac{a}{p}\right) = 1$ and because you know that $a \equiv (g^h)^2 \pmod{p}$. If k is odd then $\left(\frac{a}{p}\right) = -1$.

Eric. Here you are, back to primitive roots modulo p.

P.P. The determination of g is usually time consuming, when g is not small. This gives the advantage to Gauss's method and shows how useful the quadratic reciprocity law can be.

Eric. Where can I read about congruences?

P.P. The original treatment appeared in the famous (and still important) book by Gauss published in 1801, entitled "*Disquisitiones Arithmeticae*". The initial chapters of the book "*Introduction to the Theory of Numbers*" by L. E. Dickson, published in 1919, contain the material we have been discussing, and includes all the proofs. Today there are many "friendly" books on elementary number theory, which can be easily found in undergraduate mathematics libraries. One I especially like is by D. M. Burton, titled "*Elementary Number Theory*".

Notes About Adrien-Marie Legendre (1752–1833)

Born in Paris, France, Legendre had a long and prominent career. He began as an assistant to Pierre-Simon Laplace at the École Militaire, from 1775 to 1780. His memoir on ballistics won the first prize in the 1782 Berlin Academy contest. Soon after submitting his paper on attraction of ellipsoids, where he introduced the now-called *Legendre functions*, he was admitted as an adjunct member of the Paris Académie. This was in 1783; in 1787, he succeeded Laplace as an associate member. As such, he participated in important projects such as the preparation of tables of trigonometric functions and of logarithms, and in the organization of a coherent system of physical units, including extensive geodesic measurements, in view of determining the exact value of the meter. Work done with the Royal Observatory in Greenwich earned him election as a fellow of the Royal Society of London. Due to the French Revolution, the Academy was closed in 1791 and reopened in 1795, when Legendre became a full member. In number theory, Legendre published a long memoir in 1785 which was included in his book of 1798. After the publication in 1801 of the *"Disquisitiones Arithmeticae"* by Gauss, Legendre published a much revised second edition of his book. A third edition in 1830 was by that time considered much out of date without a substantial account of the progress in the intervening years.

Legendre stated, but couldn't prove, the quadratic reciprocity law, which was first proven by Gauss without any reference to Legendre, which caused bitter feelings on the part of Legendre. The evaluations of $\pi(N)$, the number of primes less or equal to N, given by Legendre turned out to be incorrect; this time, Legendre ignored the asymptotic conjectured value $\pi(N) \sim \frac{N}{\log N}$ which was proposed by Gauss, and almost a century later proved the so-called Prime Number Theorem.

Legendre did major pioneering work on elliptic integrals, publishing a three-volume treatise on elliptic functions. His book *"Elements de Géométrie"* replaced the classic Euclid's *"Elements"* in the teaching of elementary geometry for over 100 years in a multitude of editions and translations. For Legendre, his work was his life. This is how he wanted to be remembered.

12

Homework Checked

Paulo and Eric came with their homework. From their looks I could imagine that they were thinking: "Did I do it right? Did I make an odd number of errors, or a non-zero but even number of errors?"

I saw the pages of calculations neatly displayed, and asked...

P.P. Let me see the

First Legendre symbol:

$$\left(\frac{25432}{8227}\right) = \left(\frac{751}{8227}\right)$$

because

$$25432 = 3 \times 8227 + 751$$
$$751 = 8227 - 7476$$

so

$$\left(\frac{751}{8227}\right) = \left(\frac{8227 - 7476}{8227}\right) = \left(\frac{-1}{8227}\right)\left(\frac{7476}{8227}\right)$$

7476	2
3738	2
1869	3
623	7
89	89
1	

so
$$7476 = 2^2 \times 3 \times 7 \times 89$$
$$\left(\frac{-1}{8227}\right)\left(\frac{7476}{8227}\right) = \left(\frac{-1}{8227}\right)\left(\frac{2^2 \times 3 \times 7 \times 89}{8227}\right)$$
$$\times \left(\frac{-1}{8227}\right)\left(\frac{3}{8227}\right)\left(\frac{7}{8227}\right)\left(\frac{89}{8227}\right)$$
$$8227 \equiv 3 \pmod{4} \text{ so } \left(\frac{-1}{8227}\right) = -1.$$

By the quadratic reciprocity law

$$\left(\frac{3}{8227}\right) = -\left(\frac{8227}{3}\right) = -\left(\frac{3 \times 2742 + 1}{3}\right) = -\left(\frac{1}{3}\right) = -1.$$

$7 \equiv 3 \pmod{4}$, so

$$\left(\frac{7}{8227}\right) = -\left(\frac{8227}{7}\right) = -\left(\frac{7 \times 1175 + 2}{7}\right) = -\left(\frac{2}{7}\right) = -1.$$

89 is also a prime, $89 \equiv 1 \pmod{4}$ so

$$\left(\frac{89}{8227}\right) = \left(\frac{8227}{89}\right) = \left(\frac{89 \times 92 + 39}{89}\right) = \left(\frac{39}{89}\right) = \left(\frac{3}{89}\right)\left(\frac{13}{89}\right)$$

$$\left(\frac{3}{89}\right) = \left(\frac{89}{3}\right) = \left(\frac{3 \times 29 + 2}{3}\right) = \left(\frac{2}{3}\right) = -1$$

$$\left(\frac{13}{89}\right) = \left(\frac{89}{13}\right) = \left(\frac{13 \times 6 + 11}{13}\right) = \left(\frac{11}{13}\right)$$

$$= \left(\frac{13}{11}\right) = \left(\frac{11 \times 1 + 2}{11}\right) = \left(\frac{2}{11}\right) = -1.$$

Hence $\left(\frac{89}{8227}\right) = (-1) \times (-1) = 1.$

Now we put it all together and get

$$\left(\frac{25432}{8227}\right) = (-1) \times (-1) \times (-1) \times 1 = -1.$$

Therefore 25432 is not a square modulo 8227.

Second Legendre symbol:

$$\left(\frac{-80022}{9973}\right) = \left(\frac{-1}{9973}\right)\left(\frac{80022}{9973}\right)$$

$9973 \equiv 1 \pmod{4}$, so

$$\left(\frac{-1}{9973}\right) = 1$$

$$\left(\frac{80022}{9973}\right) = \left(\frac{9973 \times 8 + 238}{9973}\right)$$

$$= \left(\frac{238}{9973}\right) = \left(\frac{2 \times 7 \times 17}{9973}\right)$$

$$= \left(\frac{2}{9973}\right)\left(\frac{7}{9973}\right)\left(\frac{17}{9973}\right)$$

$9973 \equiv 5 \pmod{8}$ so $\left(\frac{2}{9973}\right) = -1$.

$$\left(\frac{7}{9973}\right) = \left(\frac{9973}{7}\right) = \left(\frac{7 \times 1424 + 5}{7}\right)$$

$$= \left(\frac{5}{7}\right) = \left(\frac{7}{5}\right) = \left(\frac{5 \times 1 + 2}{5}\right) = \left(\frac{2}{5}\right) = -1,$$

$$\left(\frac{17}{9973}\right) = \left(\frac{9973}{17}\right) = \left(\frac{17 \times 586 + 11}{17}\right)$$

$$\left(\frac{11}{17}\right) = \left(\frac{17}{11}\right) = \left(\frac{11 \times 1 + 6}{11}\right)$$

$$\left(\frac{6}{11}\right) = \left(\frac{2}{11}\right)\left(\frac{3}{11}\right) = (-1)(-1)\left(\frac{11}{3}\right)$$

$$= \left(\frac{3 \times 3 + 2}{3}\right) = \left(\frac{2}{3}\right) = -1.$$

Putting it together

$$\left(\frac{-80022}{9973}\right) = (-1) \times (-1) \times (-1) = -1$$

so -80022 is not a square modulo 9973.

Third Legendre symbol:

$\left(\frac{4577}{5309}\right)$. We do not need to know if 4577 is prime or composite, but we apply the trick anyway

$$\left(\frac{4577}{5309}\right) = \left(\frac{5309-732}{5309}\right) = \left(\frac{-732}{5309}\right) = \left(\frac{-1}{5309}\right)\left(\frac{732}{5309}\right).$$

From $5309 \equiv 1 \pmod 4$ we obtain $\left(\frac{-1}{5309}\right) = 1$.

We have $732 = 2^2 \times 3 \times 61$ so

$$\left(\frac{732}{5309}\right) = \left(\frac{3}{5309}\right) \times \left(\frac{61}{5309}\right) \pmod 8$$

$$\left(\frac{3}{5309}\right) = \left(\frac{5309}{3}\right) = \left(\frac{3 \times 1769 + 2}{3}\right) = \left(\frac{2}{3}\right) = -1,$$

$$\left(\frac{61}{5309}\right) = \left(\frac{5309}{61}\right) = \left(\frac{61 \times 8 + 51}{61}\right)$$

$$= \left(\frac{51}{61}\right) = \left(\frac{3 \times 17}{61}\right) = \left(\frac{3}{61}\right) \times \left(\frac{17}{61}\right).$$

But $\left(\frac{3}{61}\right) = \left(\frac{61}{3}\right) = \left(\frac{3 \times 20 + 1}{3}\right) = \left(\frac{1}{3}\right) = 1$ and

$$\left(\frac{17}{61}\right) = \left(\frac{61}{17}\right) = \left(\frac{17 \times 3 + 10}{17}\right) = \left(\frac{10}{17}\right)$$

$$\left(\frac{2}{17}\right) \times \left(\frac{5}{7}\right) = 1 \times \left(\frac{7}{5}\right) = \left(\frac{5 \times 1 + 2}{5}\right) = \left(\frac{2}{5}\right) = -1.$$

Putting it all together

$$\left(\frac{4577}{5309}\right) = 1 \times (-1) \times 1 \times (-1) = 1.$$

Hence 4577 is a square modulo 5309.

P.P. The calculations were correct! I hope you'll remember all your lives how to calculate Legendre symbols.

13

Testing for Primality and Factorization

Paulo and Eric were all ears when I announced that we were going to discuss primality testing and factorization.

P.P. Suppose you are not feeling well and you go to the doctor. He says: "I will runs some tests." When the results come back, he may declare that your health is good; he did not find any bacteria or any chemical imbalance in your blood. Or he may say that bacteria, toxins or other conditions are present, and he declares that you are sick.

Paulo. Papa Paulo, is this the primality testing you announced?

P.P. Not quite, but it is similar. You (like a doctor) want to test a number (like a patient). You found no factors (like bacteria) and you say that the number is a prime (you are healthy). If you find factors (bacteria, toxins), the number is composite (you are sick).

Paulo. It is good to know that prime numbers are healthy numbers.

P.P. Sort of. With any natural number n, our question is to find out if that number is prime or composite. As I said earlier, this is possible to do with a finite number of operations. We have to check, one after the other, if any number k is a factor of n — stopping at the smallest k such that $k^2 \geq n$. The problem is the time required to complete the testing. So what we need is an efficient method of testing.

Suppose you work with a method to test the primality of a number n. It is clear that the expected number of bit operations is bigger to test m than to test n, when m has more digits than n. One may, as I once said, estimate the number of bit operations needed to deal with n. As we take larger and larger numbers n, the number of required bit operations may be proportional to the number of digits of n; or to the square of the number of digits of n; or to a fixed power of the number of digits of n. In all these cases, we say that the *testing runs in polynomial time*. Such tests can be implemented using computers to produce the results in a reasonable time. Some testing methods do not run in polynomial time, so they are not useful for testing large numbers n.

But I have to stress that different numbers may react differently to testing. For example, if I present to you a number with 300 digits ending with 2, it is obvious that n is composite. But if I subtract 1, the new number may require much time to test for primality.

When a number n is submitted to a primality test the result (also called "the output") may be "n is a prime" or "n is composite" — but quite often, no factor of n is explicitly given.

To find factors of a large number may be quite time consuming; remember I have indicated the case of a number n which is the product of two large primes (not revealed to the person testing the number). The opposite situation are the numbers which are products of many small primes.

Eric. What are "large primes" and "many small primes"?

P.P. This is no exact concept. It is something that is dictated by available computer capability. If I have no computer at my disposal and I pick a number n with 12 digits at random, this number is much too large to treat — unless it is a friendly number with small factors that are easy to detect. At this time, a number with 150 or more digits can be treated by present day computers. I will come back to this point someday. In fact, very friendly numbers with several million digits have been tested and declared to be primes.

Papa Paulo declared: I am a friendly person, so I like friendly numbers!

P.P. Bankers are also friendly persons, but they like and must even pay to have the most unfriendly numbers.

Eric. Bankers friendly and unfriendly numbers, what is that all about?

P.P. Stick with me. I will reveal all.

P.P. Time for your soups — do you eat salads too? Let us meet again in two days at 4 p.m., OK?

14

Fermat Numbers are Friendly. Are They Primes?

Two days later...

P.P. If a number $N > 1$ is such that the prime factors of $N - 1$ are known, then N is a friendly number. There are primality tests which may be efficiently applied to N.

The easiest such numbers are of the form $N =$ (power of 2) $+ 1$.

The first question one may ask is whether a number $N = 2^m + 1$ can ever be a prime?

It is good to begin with a simple example. Let $a > 1$ and look at these products:

$$(a+1)(a^2 - a + 1) = (a^3 - a^2 + a) + (a^2 - a + 1)$$
$$= a^3 + 1,$$
$$(a+1)(a^4 - a^3 + a^2 - a + 1) = (a^5 - a^4 + a^3 - a^2 + a)$$
$$+ (a^4 - a^3 + a^2 - a + 1)$$
$$= a^5 + 1.$$

Can you guess how I continue?

Let $k > 2$ be odd. Then

$$(a+1)(a^{k-1} - a^{k-2} + a^{k-3} - a^{k-4} + \cdots - a + 1)$$
$$= (a^k - a^{k-1} + a^{k-2} - a^{k-3} + \cdots - a^2 + a)$$
$$+ (a^{k-1} - a^{k-2} + a^{k-3} - \cdots - a + 1)$$
$$= a^k + 1.$$

Eric. Once you made a similar calculation. I understand that k has to be odd, but a may be any integer greater than 1.

P.P. Yes, and I'll use your remark to prove the following lemma:

Lemma. *If $N = 2^m + 1$ is a prime then m is a power of 2.*

Proof: The number m may be equal to $1 = 2^0$, giving $2^m + 1 = 3$. Now we assume that $m \geq 2$. Say that 2^n is the highest power of 2 dividing m; I am not excluding that $n = 0$, that is $2^n = 1$, which means that m is odd. Anyway, I may write $m = 2^n \times k$ where k is odd. If k is equal to 1, we have m equal to a power of 2, as desired.

Let $k > 1$. Let $a = 2^{2^n}$ so $m = 2^{2^n \times k} + 1 = a^k + 1$. But remember, I have indicated that $a^k + 1$ has the factor $a + 1$, $1 < a + 1 < a^k + 1$ so $m = a^k + 1$ is not a prime. The conclusion is that if m is a prime then $m = 2^{2^n} + 1$, where $n \geq 0$. q.e.d.

Paulo. Are you also going to say that for every $n \geq 0$ the number $2^{2^n} + 1$ is a prime?

P.P. Why don't we examine it together? The numbers $2^{2^n} + 1$ are called *Fermat numbers*, because Fermat studied these numbers. The notation is $F_n = 2^{2^n} + 1$.

Let us make $n = 0, 1, 2, \ldots$ and see what we get.

$$F_0 = 2^{2^0} + 1 = 2 + 1 = 3 \text{ prime}$$
$$F_1 = 2^{2^2} + 1 = 2^4 + 1 = 17 \text{ prime}$$
$$F_3 = 2^{2^3} + 1 = 2^8 + 1 = 257 \text{ prime}$$
$$F_4 = 2^{2^4} + 1 = 2^{16} + 1 = 65537.$$

Fermat took pains to show that F_4 is also a prime.

Paulo. So I assume that he made the only intelligent guess possible; for every $n \geq 5$ F_n is also a prime.

Eric. I don't believe he had much evidence, just up to F_4. What happened with F_5?

P.P. Fermat calculated the number F_5; he found it had 10 digits. He needed to test if any prime up to 100 000 would be a factor of F_5.

But in his time, no such tables existed, so Fermat could not check if F_5 is a prime, and for F_n (with $n > 5$) this would even be more forbidding.

Eric. Did anyone solve or make progress in the problem during Fermat's lifetime?

P.P. No. When Fermat died he believed firmly that all Fermat numbers are primes. As he knocked at the Doors of Heaven to find if he would be directed to Heaven, Purgatory or Hell, Saint Peter examined Fermat:

> *Saint Peter*: As you know, I keep a watch over all mortals. Not only do I have a magic Third Eye, but here in Heaven, we are much more advanced than you. Digital cameras, electronic chips on everyone, GPS at work — and you poor mortals aren't even aware of all this.

[I must add that intelligent mortals have now discovered those special devices, but how about other ones that Heaven must have? Get to work, smart mortals!]

> *Saint Peter*: Fermat, you have been a magistrate, never corrupted, and a just man of law. If this was all, I would send you to Heaven. But you have been an amateur mathematician with new inventions and new ideas — integral calculus, geometry, probability and an extraordinary passion for numbers. Another strong reason to send you to Heaven. But I am sorry to say that my Third Eye caught you lying on two occasions. Remember when you proclaimed something about powers? Wait one minute, I have it here in my files. You wrote in Latin, but here we catch lies in all languages.
>
> *"Cubum autem in duos cubos, aut quadratoquadratum in duos quadratoquadratos, et generaliter nullam in infinitum ultra quadratum potestatem in duas ejusdem nominis fas est dividere: cujus rei demonstrationem*

mirabilem sane detexi. Hane marginis exiguitas nun caperet."

Saint Peter (patiently translating the Latin text to his assistant): It is impossible to separate a cube into two cubes, or a biquadrate into two biquadrates, or in general any power higher than the second into power of like degree; I have discovered a truly miraculous proof, which this margin is too small to contain.

We know that you only had the proof when $n = 4$, you maybe had a good feeling for $n = 3$; why did you not write it down? And the story of a "short margin" is poor support for your lie. The margins of that book were pretty wide. So this story is a dark spot for you. Besides, without any good reason — you only knew that F_0, F_1, F_2, F_3 and F_4 are primes — you were telling everyone that F_n is a prime for infinitely many n in the hope that the veracity of the assertion could not be contradicted. Up here, we find that you acted in bad faith. You do not deserve Paradise until the truth of what you asserted will be proved in a future generation. But you were too good to go to Hell. So it is Purgatory for you.

Eric. Today we know that "Fermat's Last Theorem" was proven true. The community of mathematicians is solidly with Fermat. Imagine, it was an English mathematician that came to his rescue.

Paulo. I know. His name is Andrew Wiles; I bet he will go to Heaven. Has anyone succeeded in showing that there are infinitely many Fermat numbers which are primes? Are they all primes?

P.P. That is our subject today. Due to the size of Fermat numbers, it was necessary to be clever. Euler threw the first stone at Fermat. Euler proved:

Theorem. *If a prime p divides the Fermat number F_n, then p is a number of the form $k \times 2^{n+2} + 1$.*

Paulo. Is this theorem of any value?

P.P. Suppose you think that F_5 is not a prime and you search for a prime factor p. You need only to test the integers which are primes among $1 \times 2^7 + 1 = 129$, $2 \times 2^7 + 1 = 257$, etc., so your work will be 128 times less. This is good for lazy people. Now, the theorem does not tell that such a prime p exists which is smaller than $2^{2^n} + 1 = 2^{2^n - n - 2} \times 2^{n+2} + 1$. To succeed you would have to try many primes.

The proof is quite nice, so I will give it to you. It uses several facts which we already learned about congruences.

Proof of the theorem: Suppose that p divides $2^{2^n} + 1$, so $2^{2^n} + 1 \equiv 0 \pmod{p}$, hence $2^{2^n} \equiv -1 \pmod{p}$. Raising to the square, $2^{2^{n+1}} \equiv 1 \pmod{p}$. We just saw that 2^{n+1} is the order of 2 modulo p, because this order e has to divide 2^{n+1} and it is not 2^n, nor any smaller power of 2. We note that $n + 1 \geq 3$, so 8 divides 2^{n+1}. Euler used Fermat's theorem: $2^{p-1} \equiv 1 \pmod{p}$, hence 2^{n+1} divides $p - 1$. Combining, 8 divides $p - 1$, that is $p \equiv 1 \pmod{8}$, remember, 2 is a square modulo p, hence $1 = \left(\frac{2}{p}\right)$. Remember also that we proved that $\left(\frac{2}{p}\right) \equiv 2^{\frac{p-1}{2}} \pmod{p}$. Then the order 2^{n+1} of 2 modulo p, divides also $\frac{p-1}{2}$.

Hence raising to the square 2^{n+2} divides $p - 1$, therefore there exists a natural number k such that $p - 1 = k \times 2^{n+2}$ and $p = k \times 2^{n+2} + 1$.
$$\text{q.e.d.}$$

Paulo. Euler knew how to go around. I don't feel that I could do it.

Eric. Maybe Euler spent a long time in doing it. One year, maybe.

P.P. Frankly, I don't think so. Euler had so many discoveries, almost two meters of theorems (I measured them). He was very smart.

Paulo. I didn't know that one measures number of theorems by meters. How is this possible?

P.P. I went to the library, which has (if I remember correctly) 39 volumes of a still incomplete series of books containing Euler's theorems. They measured about two meters. (*As I said it, I realized that I was exaggerating, two meters would mean volumes over*

25 cm thick; but I left it as I said). But Euler was lucky with F_5. When he took the prime $641 = 5 \times 2^7 + 1$, he found that
$$F_5 = 641 \times 6700417$$
so F_5 is not a prime. This was sheer luck. The fact is that it is more difficult to search for actual factors of F_n than to submit F_n to a primality test with the possible output "prime" or "composite", without any indication of a prime factor in the second alternative. So I will give a primality test that is related to Fermat's little theorem.

Remember ...?

Paulo. I do, if p is a prime and $1 < a < p$ then $a^{p-1} \equiv 1 \pmod{p}$.

P.P. Can I say something? It is possible to have a number $n > 2$, which is composite, but for which there exists a number
$$a,\ 1 < a < n \quad \text{with} \quad \gcd(a, n) = 1 \quad \text{and} \quad a^{n-1} \equiv 1 \pmod{n}.$$

Eric. A little bit like if n would be a prime. Can you show me an odd number n such that $2^{n-1} \equiv 1 \pmod{n}$?

P.P. Today I can, but for a long time these numbers were unknown. Once you find it, it is easy to check. Take $n = 11 \times 31 = 341$. You need not work much to find that $2^{340} \equiv 1 \pmod{341}$.

Paulo. Maybe there are only a few such numbers.

P.P. On the contrary. There are infinitely many composite numbers n such that $2^{n-1} \equiv 1 \pmod{n}$. Of these, 341 is the smallest. Such numbers are called *pseudoprimes for the basis* 2.

Eric. How about if instead of 2 you choose a number $b > 2$?

P.P. If $\gcd(b, n) = 1$ and $b^{n-1} \equiv 1 \pmod{n}$, then n is called a *pseudoprime for the basis* b. Before you ask, $91 = 7 \times 13$ is the smallest pseudoprime for the basis 3. Today it has been proven that for any number $b \geq 2$ there exists infinitely many pseudoprimes for the basis b.

Eric. What an annoyance with respect to Fermat's little theorem. A number n is composite and there is a number b such that $1 < b < n$, $\gcd(b, n) = 1$ and yet $b^{n-1} \equiv 1 \pmod{n}$. You know what I think?

P.P. What do you think?

Eric. The world is wrong; this should be forbidden.

P.P. Well, how about this one? There exists a natural number $n > 2$ such that first: n is composite, and second: (watch what I say) for *all* integers a, $1 < a < n$, with $\gcd(a,n) = 1$, we have $a^{n-1} \equiv 1 \pmod{n}$.

Eric. I cannot believe it. This is a provocation. If this is true, show me one.

P.P. $561 = 3 \times 11 \times 17$. It takes a little bit more time to check, but you'll be able to do it.

Eric. Amazing! This surprises me. It must be very exceptional. Do you know the largest number with this property? Do these bad numbers have a name?

P.P. A name they do have, not from the discoverer, but from Carmichael, who studied them. The actual discoverer was Korselt, but few people know it.

Paulo. Ah! OK! It is like Columbus who came to the continent we call America, long after Eric the Red came with the Vikings.

Eric. Maybe Eric the Red was my ancestor. Maybe this explains why I feel so well in America, which I will now call Ericsland. But Papa Paulo, which is the largest Carmichael number?

P.P. There isn't one. Three smarties all of whom will go to Heaven, proved that there are infinitely many Carmichael numbers.

Paulo. Tell me their names. If I meet them, I can ask for an autograph.

P.P. "Red" Alford is already in Heaven. Granville and Pomerance can be found in meetings of the most sophisticated level. If you ask for an autograph, tell them that you know Papa Paulo — then they will not refuse you.

Eric. Infinitely many bad Carmichael numbers. So a primality test involving the ideas of Fermat's little theorem must be impossible.

P.P. Not really, just some precaution is needed. Lucas, and later Lehmer, had a safe method. I give now a primality test which is the appropriate converse of Fermat's little theorem.

Primality test. Let n be an odd integer, $n > 1$. We assume that for every prime q dividing $n-1$, there exists an integer a (this a depends on q, so if q' is another prime dividing $n-1$, the corresponding a' need not be equal to a, but it could be equal) such that $1 < a < n$ and $a^{n-1} \equiv 1 \pmod{n}$, while $a^{\frac{n-1}{q}} \not\equiv 1 \pmod{n}$. Then n is a prime.

The proof is fully accessible to both of you. I think that the best thing would be for you to help me, while I conduct the proof. First question: do you remember Euler's function $\varphi(n)$ and its meaning?

Paulo. I do not forget important things like Euler's function: $\varphi(n)$ is the number of integers a such that $1 \leq a < n$ and $\gcd(a, n) = 1$.

P.P. Good. Which number is bigger? $\varphi(n)$ or $n - 1$?

Paulo. Clearly $\varphi(n)$ is at most equal to $n - 1$. But it may be equal.

P.P. Your last sentence is important. If $\varphi(n) = n - 1$ it means that every a such that $1 \leq a < n$ is coprime with n.

Eric. So if $1 < a < n$, then a does not divide n. So n is a prime.

P.P. Excellent, so we have to show that $\varphi(n) = n - 1$. My strategy is to show that $n - 1$ divides $\varphi(n)$. If I have shown it, then of course $n - 1 \leq \varphi(n)$. But Paulo said that $\varphi(n) \leq n - 1$, therefore $\varphi(n) = n - 1$.

Paulo. Your task is to show that $n - 1$ divides $\varphi(n)$. But I wonder how you can do it, since you don't know which is the number n; it is unspecified.

P.P. That is a great remark. This tells me to try reduction to an absurd. So I shall assume that $n - 1$ does not divide $\varphi(n)$.

Here is a good use of Euclid's theorem. I write $n - 1 =$ (product of powers of primes, like q^r) and $\varphi(n) =$ (product of powers of primes like q^s). Remember, $n - 1$ does not divide $\varphi(n)$, so there must be some q^r (with $r \geq 1$) which divides $n - 1$ but does not divide $\varphi(n)$. And associated to the prime q, which divides $n - 1$, by assumption there exists an integer a, $1 < a < n$, a coprime to n, satisfying $a^{n-1} \equiv 1 \pmod{q}$, $a^{\frac{n-1}{q}} \not\equiv 1 \pmod{n}$.

Paulo. Not so fast. What power of q divides $n-1$?

Eric. Remember. There exists the smallest e, $1 < e$, such that $a^e \equiv 1 \pmod{n}$.

Paulo. And e divides $n-1$, but e does not divide $\frac{n-1}{q}$.

P.P. Question: what is the power of q dividing e?

Paulo. Since e divides $n-1$, the power is no more than q^r. But e does not divide $\frac{n-1}{q}$, so it cannot be less than q^r.

Eric. Why?

Paulo. I will explain from scratch. I write $e = q^s e'$ where q does not divide e' and $0 \le s$. For $n-1 = q^r m$, where q does not divide m and from e dividing $n-1$, then $s \le r$. We have $\frac{n-1}{q} = q^{r-1} m$ and e does not divide $\frac{n-1}{q}$, so s is not less or equal to $r-1$, so $s = r$.

P.P. I will finish the proof. By Euler's theorem $a^{\varphi(n)} \equiv 1 \pmod{n}$, hence also e divides $\varphi(n)$ and finally q^r divides $\varphi(n)$.

Eric. Stop Papa Paulo! You played your favorite game, "Contradiction", so $n-1$ divides $\varphi(n)$, so $n-1 \le \varphi(n)$, so $n-1 = \varphi(n)$, so n is a prime. The proof is finished. q.e.d.

P.P. This primality test is good for friendly numbers n. You must know all the prime factors of $n-1$. So it may, and has been applied to test the primality of Fermat numbers $F_n = 2^{2^n} + 1$.

Paulo (who shies away from extra work) had a pressing question:

Paulo. For large n, you may have to try several integers a, to see if the conditions are satisfied. For large n, there are quite a lot, in fact $\frac{n-1}{q}$ powers of a to be calculated modulo n. What a job!

P.P. It is possible to cut this work with a trick. I will give you an example with numbers. I'll show how to quickly calculate 3^{55}

(mod 36). First write 55 in base 2, that is $55 = 2^5 + 2^4 + 2^2 + 2 + 1$. So $3^{55} = 3^{2^5} \times 3^{2^4} \times 3^{2^2} \times 3^2 \times 3$. We obtain, successively by squaring,

$$3^2 \equiv 9 \pmod{36}$$
$$3^4 \equiv 81 \equiv 9 \pmod{36}$$
$$3^8 \equiv 81 \equiv 9 \pmod{36}$$
$$3^{16} \equiv 81 \equiv 9 \pmod{36}$$
$$3^{32} \equiv 81 \equiv 9 \pmod{36}$$

then
$$3^{55} \equiv 9 \times 9 \times 9 \times 9 \times 3 \equiv 27 \pmod{36}.$$

Paulo. I got it. You did not explain how to write a number in base 2, but that is easy to guess. I will demonstrate with 100: The highest power of 2 not more than 100 is $2^6 = 64$. I subtract: $100 - 2^6 = 36$. The highest power of 2 not greater than 36 is $2^5 = 32$. I subtract $36 - 2^5 = 4 = 2^2$. Putting together $100 = 2^6 + 2^5 + 2^2$.

P.P. It is good to talk with intelligent people! So my trick reduces by far the number of operations to obtain $a^{\frac{n-1}{2}} \pmod{n}$ and $a^{\frac{n-1}{q}} \pmod{n}$.

Papa Paulo continued:

P.P. Fermat numbers $F_n = 2^{2^n} + 1$ are the friendliest numbers, to apply the primality test. Here is what we get:

Primality test for Fermat numbers. Let $n > 2$ and assume that $3^{\frac{F_n-1}{2}} \equiv -1 \pmod{F_n}$. Then F_n is prime.

No need for a proof. Just note that $F_n - 1 = 2^{2^n}$, so $\gcd(3, 2^n) = 1$. We may apply directly the previous primality test to conclude that F_n is prime.

Eric. OK. And what can we say if, after all the work $3^{\frac{F_n-1}{2}} \not\equiv -1 \pmod{F_n}$?
It would be good if one could conclude that F_n is composite.

P.P. This is so. I repeat: Let n be an integer such that $n > 1$. If $3^{\frac{F_n-1}{2}} \not\equiv -1 \pmod{F_n}$ then F_n is composite. For this, I need a proof.

Proof: I will assume that F_n is a prime.

We have $F_n = 2^{2^n} + 1 \equiv 1 \pmod 4$ and $F_n = 2^{2^n} + 1 \equiv 1 + 1 \equiv 2 \pmod 3$.

So $\left(\dfrac{3}{F_n}\right) = \left(\dfrac{F_n}{3}\right) = \left(\dfrac{2}{3}\right) = -1.$

Here I have used the quadratic reciprocity law. Another property of the Legendre symbol is

$\left(\dfrac{3}{F_n}\right) \equiv 3^{\frac{F_n-1}{2}} \pmod{F_n}$. Hence $3^{\frac{F_n-1}{2}} \equiv -1 \pmod{F_n}$.

This says that if $3^{\frac{F_n-1}{2}} \not\equiv -1 \pmod{F_n}$, then F_n has to be composite.

q.e.d.

Paulo. Now it is clear what to do. Apply the test to F_6 to get the output. Do the same with F_7, F_8 and continue as long as you want. Do you know what has been found so far?

P.P. I can only say what I know today. As more becomes known, you will have to update this information.

Paulo. Has any Fermat number F_n with $n > 4$ been certified to be prime?

P.P. All F_n already tested are composite!

Eric. For which F_n besides F_5 is the factorization known?

P.P. The complete factorization is known for F_5 (as said earlier) and for $F_6, F_7, F_8, F_9, F_{10}$ and F_{11}. For example F_{11} is the product of 5 distinct primes, of which the largest one has 564 digits. It was a computational feat to show that this factor was indeed a prime.

On the other side, the factorization of F_6 was indicated by Clausen in a letter to Gauss. Unaware, in 1880 (before computers were invented) Landry found that $F_6 = 274177 \times 67280421310721$. It took him three years of Sundays to perform the factorization.

Paulo. I am impressed by Landry's patience. Now, are there any Fermat numbers with incomplete factorizations?

P.P. Of course, because you hit large factors which you cannot decide if they are primes, since they are unfriendly. So the factorization of F_{12}, F_{13}, F_{15}, F_{16}, F_{17}, F_{18}, F_{19}, F_{21} and F_{23} are incomplete.

Eric. You missed a few Fermat numbers.

P.P. F_{14}, F_{20}, F_{22} and F_{24} are composite but no factor of these numbers is known.

Paulo and Eric were impressed by the state of incomplete knowledge concerning Fermat numbers. I added:

P.P. Presently it is not known if there are infinitely many composite Fermat numbers and it is also not known if there are infinitely many prime Fermat numbers.

The effect of this strong statement of ignorance caused this reaction on Paulo and Eric: Poor Fermat, he may stay in purgatory forever.

15

This World is Perfect

P.P. Perfect numbers were already known in ancient times. The first perfect number was 6; it was connected by mystics and religious writers to perfection. Creation required 6 days, so PERFECT is this world.

The next perfect number is 28, which is roughly the number of days the Moon needs to orbit the Earth.

Paulo. Is the next perfect number equal to the number of Saturn days that Titan needs to orbit Saturn?

P.P. That is something I really don't know. Perfect numbers are not defined in biblical or astronomical terms. They have a simple arithmetical definition:

Look at the divisors of 6 smaller than 6. They are 1, 2 and 3. Their sum is $1 + 2 + 3 = 6$.

Now look at the divisors of 28 smaller than 28: 1, 2, 4, 7, and 14. Their sum is $1 + 2 + 4 + 7 + 14 = 28$.

A *perfect number* is a natural number n such that the sum of the divisors of n smaller than n are equal to n. It is pretty easy to check. Not only 6 and 28, but also 496 and 8128 are also perfect numbers. In fact, these are the only perfect numbers less than 10 000.

Eric. Of course, it is easy to check that 496 and 8128 are perfect numbers. But to find that there are no others less than 10 000 requires checking each $n < 10\,000$ to find its smaller divisors and compare their sum with n. Quite a bit of work.

Paulo. What were the mathematicians doing?

P.P. Our old friend Euclid already discovered a method to produce perfect numbers. To be exact, Euclid proved that if "something happens" for the integer k, then a formula gives a perfect number.

Paulo. What is "something happens"?

P.P. Just wait; I will make it clear in due time, but first let me say ...

Eric interrupted:

Eric. I know what you are going to say. My question would have been: Can one have perfect numbers which are not defined by Euclid's prescription?

P.P. Yes, that was going to be my comment. All will become clear once I state and prove Euclid's theorem.

Euclid's Theorem for Perfect Numbers. *Let $k > 1$ be an integer such that $2^k - 1$ is a prime. Then $P_k = 2^{k-1}(2^k - 1)$ is a perfect number.*

Thus, if $k = 2, 3, 5$ or 7, we have

$$P_2 = 2(2^2 - 1) = 6 \text{ perfect}$$
$$P_3 = 2^2(2^3 - 1) = 28 \text{ perfect}$$
$$P_5 = 2^4(2^5 - 1) = 496 \text{ perfect}$$
$$P_7 = 2^6(2^7 - 1) = 8128 \text{ perfect}$$

because $2^2 - 1$, $2^3 - 1$, $2^5 - 1$ and $2^7 - 1 = 127$ are primes.

Paulo. Wonderful. You need to know when $2^k - 1$ is a prime.

P.P. This is one very important problem which I will discuss later. For the moment, I will show that it does happen that $2^k - 1$ is not

a prime:
$$2^4 - 1 = 15,$$
$$2^6 - 1 = 63,$$
$$2^8 - 1 = 255,$$
$$2^9 - 1 = 511 = 7 \times 73,$$
$$2^{10} - 1 = 1023 = 3 \times 11 \times 31,$$
$$2^{11} - 1 = 2047 = 23 \times 89.$$

Paulo. Enough. I already see that to decide when $2^k - 1$ is a prime will be a tough question. Are you going to prove Euclid's theorem? He is one of my favorite characters, you know. We already saw a few of his theorems that are very important and that were discovered so long ago.

P.P. Yes, I will now give the proof.

Proof: Let $2^k - 1$ be a prime. The divisors of $2^k - 1$ are 1 and $2^k - 1$. On the other hand, the divisors of 2^{k-1} are $1, 2, 2^2, \ldots, 2^{k-1}$. So the divisors of $2^{k-1}(2^k - 1)$ are $1, 2, 2^2, \ldots, 2^{k-1}$ as well as $2^k - 1$, $2 \times (2^k - 1), 2^2 \times (2^k - 1), \ldots, 2^{k-1} \times (2^k - 1)$. The sum of the divisors is $(1 + 2 + 2^2 + \cdots + 2^{k-1}) + ((2^k - 1) + 2 \times (2^k - 1) + 2^2 \times (2^k - 1) + \cdots + 2^{k-1}(2^k - 1)) = (1 + 2 + \cdots + 2^{k-1}) \times (1 + (2^k - 1)) = (1 + 2 + 2^2 + \cdots + 2^{k-1}) \times 2^k$.

But remember that $1 + 2 + 2^2 + \cdots + 2^{k-1} = \frac{2^k - 1}{2 - 1} = 2^k - 1$. The required sum of divisors is $(2^k - 1) \times 2^k = 2 \times 2^{k-1}(2^k - 1)$. This shows that $2^{k-1}(2^k - 1)$ is a perfect number, because the sum of smaller divisors of $2^{k-1}(2^k - 1)$ is equal to this number itself.

q.e.d.

Paulo. All the perfect numbers obtained with Euclid's prescription are even. He could not consider odd perfect numbers?

Eric. And it was not shown that each even perfect number is obtained with Euclid's prescription.

P.P. Paulo, I will address your comment later, but first I will prove Euler's theorem. That will satisfy Eric.

Euler's Theorem. *If n is an even perfect number, there exists $k > 1$ such that $n = 2^{k-1}(2^k - 1)$.*

Proof: Let n be an even perfect number. We write $n = 2^e \times m$ where $e \geq 1$ (because n is even) and m is odd. The divisors of 2^e are $1, 2, 2^2, \ldots, 2^e$, we denote by $d_0 = 1, d_1, d_2, \ldots, d_t = m$ the divisors of m.

Then the divisors of $n = 2^e m$ are the numbers $2^f d_i$, for $0 \leq f \leq e$ and $0 \leq i \leq t$. Let $s = d_0 + d_1 + \cdots + d_t$ be the sum of divisors of m. Then the sum of divisors of n is $(1 \times s) + (2 \times s) + (2^2 \times s) + \cdots + (2^e \times s) = (1 + 2 + \cdots + 2^e) \times s = (2^{e+1} - 1)s$. But n is perfect, so $(2^{e+1} - 1)s = n + n = 2n = 2^{e+1}m$. Since $2^{e+1} - 1$ is odd, then $2^{e+1} - 1$ divides m, we write $(2^{e+1} - 1)m' = m$. Therefore $(2^{e+1} - 1)s = 2^{e+1} \times (2^{e+1} - 1)m'$, so $2^{e+1}m' = s \geq m + m' = (2^{e+1} - 1)m' + m' = 2^{e+1}m'$. This implies that $s = m + m'$. But m has at least the divisor m and 1; if the divisors of m are m and m' then $m' = 1$ and so $m = 2^{e+1} - 1$ and m is a prime. So we just take $k = e - 1$ and the proof is concluded, because the sum of the smaller divisors of $2^{k-1}(2^k - 1)$ is equal to the number. q.e.d.

Paulo. This is great! I like it. The meaning is that the even perfect numbers are exactly the numbers $P_k = 2^{k-1}(2^k - 1)$ for which $2^k - 1$ is a prime, and no others.

Eric. And we have to wait for our next discussion to find when $2^k - 1$ is a prime. But I do not want to wait to learn about odd perfect numbers. Papa Paulo, can you give me some examples of odd perfect numbers?

Papa Paulo remained silent for a bit, then said...

P.P. No one knows of any odd perfect number. Perhaps there is no odd perfect number. This question is full of mystery to all except the Perfect Being, who ordered us Imperfect Mortals ...

> *Perfect Being*: Go out, you, and find if there exists an odd perfect number!

With conviction, Papa Paulo said:

P.P. In this perfect world we are mortals, all of us. Our bodies disappear into dust.

After a short interruption to give weight to his words, Papa Paulo continued.

P.P. But our work may be immortal. Euclid, Euler, Fermat and Gauss. Their names will never disappear. Nor will those of the great literary, art and music figures. Who does not know Wolfgang?

Paulo. Amadeus.

Eric. Mozart!

P.P. Yes, my friends, as an imperfect mortal, I had to satisfy the Perfect Being. And I worked hard, went to different countries, to basements of libraries, searched in dusty archives. My purpose was to find an answer to the following question: If an odd perfect number exists, which properties must it have? How big, how many prime factors, etc. The more constraints I knew, the more definite could be my search.

Eric. I think it is like police work. A criminal (that is, an odd perfect number) has to be found. Some witnesses have said: It is a man who moves fast, was dressed in dark clothes, no beard, not short, not tall. All this is very little information. Unfortunately there are no fingerprints or DNA. The work is hard.

P.P. Yes, that is about right. After all my efforts, the idea came to check that (hated) book by Miobnebir in the hope that nothing would be there. But HE HAD IT ALL! Whether I like it or not, I must concede that he did a good job of gathering the work of so many imperfect mortals in one place — many already gone, but not immortal.

Eric. And what did you find?

P.P. Suppose that N is an odd perfect number.

(a) The number of distinct prime factors of N has to be at least 8. If N is not a multiple of 3 then the number of distinct prime factors has to be at least 11.

(b) If one conceives the existence of an odd perfect number N with k distinct prime factors, then N has to be smaller than 4^{4^k}.

The numbers 4^{4^k} are very big. If one searches an odd perfect number N with 8 distinct prime factors the search would go up to $4^{4^8} = 4^{2^{16}} = 2^{2^{17}}$. We already know that this is unreachable. Should one make that effort? Maybe a more clever mortal will prove that the above number 4^{4^k} may be drastically reduced. All this means is that this type of information is insufficient.

(c) N must have at least 300 digits. And we expect that mortals will substantially increase beyond 300. N is going to be very hard to catch.

(d) N is divisible by a prime power p^e where $p \equiv 1 \pmod{4}$, $e \equiv 1 \pmod{4}$; moreover $\frac{N}{p^e}$ is a square.

(e) The largest prime factor of N must be greater than $1\,000\,000$, the second largest prime factor of N must be larger than $10\,000$, the third largest prime factor of N must be larger than 100.

Eric. If this is the kind of information you gathered, then frankly, they are no more than weak leads to find the criminal. We would need DNA.

P.P. Yes, in mathematical terms all we know is insufficient to guarantee that there exists an odd perfect number.

Paulo. A perfect crime in this perfect world. One does not even know if a criminal exists!

16

Unfriendly Numbers from a Friend of Fermat's

Back to our discussions! The day's theme was the investigation of primes of the form $2^k - 1$.

P.P. As Euclid proved, the numbers $2^{k-1}(2^k - 1)$, when $k \geq 2$ and $2^k - 1$ is prime, are even perfect numbers. Conversely, by Euler's theorem, any even perfect number is equal to $2^{k-1}(2^k - 1)$, where $k \geq 2$ and $2^k - 1$ is a prime. So there was great interest in the discovery of prime numbers of the form $2^k - 1$. As shown below, we get the following primes:

$$2^2 - 1 = 3$$
$$2^3 - 1 = 7$$
$$2^5 - 1 = 31$$
$$2^7 - 1 = 127$$
$$2^{13} - 1 = 8191$$
$$2^{17} - 1 = 131071$$
$$2^{19} - 1 = 524287.$$

The verification that $2^{13} - 1$, $2^{17} - 1$ and $2^{19} - 1$ was first done by trying if any smaller prime would divide the number in question.

On the other hand, $2^{11} - 1 = 23 \times 89$. For large exponents k the primality of $2^k - 1$ requires calculations which we shall describe.

But first let us examine a simple result which reduces the search of primes of the form $2^k - 1$.

If k is composite then $2^k - 1$ is also composite.

Proof: Let $k = nh$ where $1 < n < k$ and $1 < h < k$. Then $2^k - 1 = 2^{nh} - 1 = (2^n - 1)(2^{n(h-1)} + 2^{n(h-2)} + \cdots + 2^n + 1)$. This is true; you just have to perform the multiplication on the right-hand side. There will be cancellations and the result is $2^k - 1$. Note that the two factors of the right are greater than 1. So $2^n - 1$ is a composite integer.   q.e.d.

Paulo. Good, we shall only look at the numbers $2^q - 1$, where q is a prime. But as you pointed out, $2^{11} - 1$ is not a prime. I am inclined to think that for many primes q, $2^q - 1$ will not be a prime.

P.P. You are not wrong in your feelings. Euler proved:

Let q be a prime of the form $q > 3$ and $q \equiv 3 \pmod{4}$.

(1) If $2q + 1$ is a prime then $2q + 1$ divides $2^q - 1$.
(2) If $2q + 1$ divides $2^q - 1$ then $2q + 1$ is a prime.

Proof: (1) We assume that $p = 2q + 1$ is a prime.

From $q \equiv 3 \pmod 4$ it follows that $2q \equiv 2 \times 3 \pmod 8$, so $p = 2q + 1 \equiv 6 + 1 \equiv 7 \pmod 8$. Do you remember? This implies that 2 is a square modulo p. So there exists an integer m such that $2 \equiv m^2 \pmod p$. But $2^q = 2^{\frac{p-1}{2}} \equiv m^{2 \times \frac{n-1}{2}} \equiv m^{p-1} \equiv 1 \pmod p$ by Fermat's little theorem. So p divides $2^q - 1$.

(2) Now we assume that $m = 2q + 1$ divides $2^q - 1$. From $2^q \equiv 1 \pmod m$ then $2^{m-1} = 2^{2q} = (2^q)^2 \equiv 1 \pmod m$. We shall use the primality test which was proven earlier. The primes dividing $m - 1 = 2q$ are 2 and q. We have $2^2 \not\equiv 1 \pmod m$. Also $(-2)^{m-1} \equiv 2^{m-1} \equiv 1 \pmod m$, because m is odd, and $(-2)^q = -2^q \equiv -1 \not\equiv 1 \pmod m$. By the primality test, we conclude that $m = 2q + 1$ is a prime.   q.e.d.

Paulo. This is a very nice helper to decide if $2^q - 1$ is a prime. First I begin a list with 7, jump by 4's and take note of the primes, striking out the composite numbers

$$\begin{array}{cccccc} 7 & 11 & \cancel{15} & 19 & 23 & \cancel{27} \\ 31 & \cancel{35} & \cancel{39} & 43 & 47 & \cancel{51} \\ \cancel{55} & 59 & \cancel{63} & 67 & 71 & \cancel{75} \\ 79 & 83 & \cancel{87} & \cancel{91} & \cancel{95} & \cancel{99} \end{array}$$

and I stop here.

For each prime q which I found in the above list, I calculate $2q + 1$ and strike out the composite numbers

$$\begin{array}{cccccc} \cancel{15} & 23 & \cancel{39} & 47 & \cancel{63} & \cancel{87} \\ \cancel{95} & \cancel{119} & \cancel{135} & \cancel{143} & \cancel{159} & 167 \end{array}$$

Conclusion: $2^{11} - 1$, $2^{23} - 1$ and $2^{83} - 1$ are composite.

Eric. If you are patient you may have more primes q such that $2^q - 1$ is composite.

P.P. Once I did some calculations and found that $2^{131} - 1$, $2^{179} - 1$, $2^{191} - 1$, $2^{239} - 1$ and $2^{251} - 1$ are also composite. This is, of course, a fast way to find composite numbers $2^q - 1$. But these are exactly the ones not interesting to us. We want primes $2^q - 1$.

Eric. May I ask two questions?

P.P. Of course!

Eric. Question 1: Can I find many — infinitely many — primes q such that $q \equiv 3 \pmod{4}$?

Question 2: Can I find many — infinitely many — primes q such that $q \equiv 3 \pmod{4}$ and $2q + 1$ is also a prime?

P.P. Listen to these questions, Paulo. Both show that our Eric has acquired a great insight.

The answer to Question 1 is: There are infinitely many primes q of the form $q \equiv 3 \pmod{4}$. There is a proof of this fact which is not too hard, using only facts from elementary number theory.

The answer to Question 2 is unknown. Specialists are convinced that the problem is very difficult. Very, very difficult.

Paulo. So, where are we now? I would also like to ask my question: Is there any test for the primality of the numbers $2^q - 1$?

P.P. There had to be one. Listen to this amazing story. Mersenne was a friend of Fermat's (not friends who would go to a coffee shop or share a bottle of cognac). In fact, they met only once. But they maintained a steady correspondence, not the casual email kind, but the kind with flowery style, written with the *"plume d'oie"* dipped in black ink. I would say that they were scientific pen pals.

Paulo. I also have two pen pals, and just now a good idea. I'll not write about primes — my pals are ignorant — but about snakes and minerals!

P.P. Well, Fermat and Mersenne did not write about snakes and minerals. Fermat usually asked Mersenne to spy on other scientists, to learn their discoveries and communicate them back to Fermat. Fermat was in Toulouse, not a big intellectual center, while Mersenne was in the great Paris with direct access to the important men of his time, whether in mathematics, mechanics, physics, you name it. Usually Fermat wrote (in the beautiful style of his time): *"Cher, I discovered such and such theorem on numbers. Do propose as a challenge, the following problem to ..."* — depending on the question, it could be Pascal, Descartes, Carcavi, Wallis, Brouncker or others.

Mersenne, who was an expert in musical acoustics, had a keen interest in numbers, and he wrote back: *"Cher Fermat, the numbers $2^{13} - 1$, $2^{17} - 1$, $2^{31} - 1$, $2^{67} - 1$, $2^{127} - 1$, and $2^{257} - 1$ are primes."*

But Mersenne paid Fermat back by not revealing how he established his statement. Which amazing method had he discovered to treat such large numbers? Mersenne's statement was an enigma (for 67, 127 and 257) and mathematicians did not know how to prove what Mersenne claimed for the larger numbers. Up to $2^{19} - 1$ the assertion was easy to verify. With trial division and much patience, Euler proved that $2^{31} - 1$ is indeed a prime — the largest prime known at that time. Euler could spare many calculations, because if the integer $n > 1$ divides $2^q - 1$ with $q > 2$, then $n \equiv 1 \pmod{8}$

or $n \equiv -1 \pmod{8}$ and $n \equiv 1 \pmod{q}$. This cuts the number of required divisions.

Paulo. Can you prove what you just said?

P.P. Easily. First, I observe that if two numbers a and b are congruent to 1 or to -1 modulo 8, then the same is true for ab. In the same way, if a and b are congruent to 1 modulo q then ab is congruent to 1 modulo q. So to prove the statement it suffices to show that every prime factor p of $2^q - 1$ is congruent to 1, or to -1 modulo 8 and also congruent to 1 modulo q.

Paulo. Up to now, it is all clear. You are preparing for the proof.

P.P. OK, let p be a prime which divides $2^q - 1$, so p is odd and $2^q \equiv 1 \pmod{p}$. The order of 2 modulo p divides q, hence it is equal to q. By Fermat's little theorem $2^{p-1} \equiv 1 \pmod{p}$, hence q divides $p - 1$. We write $p - 1 = 2kq$ because $p - 1$ is even and q is odd. We calculate the Legendre symbol

$$\left(\frac{2}{p}\right) \equiv 2^{\frac{p-1}{2}} = 2^{kq} \equiv (2^q)^k \equiv 1 \pmod{p}, \text{ so } \left(\frac{2}{p}\right) = 1.$$

As it was shown once — remember? — either $p \equiv 1 \pmod{8}$ or $p \equiv 7 \equiv -1 \pmod{8}$. And that is it, or if you like,   q.e.d.

P.P. Paulo I have satisfied your request. Let me return to where I was.
Ah, yes. It was about Euler showing with calculations that $2^{31} - 1$ is a prime. More could not be done until Lucas had the idea of using recurring sequences to test the primality of the number $M_q = 2^q - 1$, which are now famous and called *Mersenne numbers*. With his new method, Lucas proved in 1878 that M_{127} is a prime. It was by far, and until the age of computers, the largest known prime.

Paulo. $2^{127} - 1$. What patience and long calculations for Lucas.

P.P. I will explain his primality test, which is straightforward to understand and quite practical.

Eric. Now I'm curious. What is this powerful test?

P.P. It is very easy to explain. Start with $S_0 = 4$, $S_1 = S_0^2 - 2 = 14$, $S_2 = S_1^2 - 2 = 194$ and go on like this. So, for every n let $S_n = S_{n-1}^2 - 2$. Then compare M_q with S_{q-2}. Lucas' test says: M_q is a prime when M_q divides S_{q-2} and M_q is composite when M_q does not divide S_{q-2}. I note that if M_q is composite, the method of testing does not give any factor of M_q.

Eric. The test is easy to understand, but it requires quite a lot of squaring when q is big. And how to prove the test?

Paulo. I bet the proof has to be of a new kind. Not like the primality test for Fermat numbers. There is the sequence of S's.

P.P. Yes, the proof is entirely different. I cannot do it because the required preparation is too long.

Eric. If it is too long for today's discussion, I suggest that you give the proof another day.

Paulo. Good idea, it is your debt to prove Lucas' test. We shall harass you like creditors until you pay!

P.P. I always pay my debts.

Eric came back to Lucas:

Eric. And how about the calculations?

P.P. For M_{127} it is quite feasible with Lucas' test. This number has 39 digits. It is obtained with 7 successive squarings:

$$2, \quad 2^2, \quad (2^2)^2 = 2^4, \quad (2^4)^2 = 2^8, \quad (2^8)^2 = 2^{16},$$

$$(2^{16})^2 = 2^{32}, \quad (2^{32})^2 = 2^{64}, \quad (2^{64})^2 = 2^{128}$$

and $M_{127} = \frac{2^{128}}{2} - 1$.

On the other hand, the calculation of the S's demands more labor, in the present case 125 squarings.

Paulo. But all you need is to find out if S_{125} is divisible by M_{127}. You can calculate least residues modulo the number M_{127} with 39 digits. OK, I concede, it is still considerable work to do.

P.P. But Lucas' test was the only reasonable way to crack Mersenne's enigma. So it was found that M_{67} is not a prime. Other Mersenne numbers, namely M_{61}, M_{89} and M_{107} were tested and certified to be primes.

Paulo. What was the fate of M_{257}?

P.P. It had to wait until 1932, when Derrick H. Lehmer, a master in calculations, showed that M_{257} is not a prime.

Eric. The End of Mersenne's enigma. Apart from the little values $q \leq 31$ (which were known), Mersenne was right for $q = 127$ and wrong for $q = 67$ and 257.

Paulo. At school, Mersenne would flunk.

P.P. All this happened before the age of computers. In 1951 the famous Alan Turing, working with early generation computers, tried to discover larger Mersenne primes. Unsuccessfully.

Eric. This made him famous?

Paulo. He is famous because of his work in cryptography during World War II. There is a book about him titled *"Codebreakers"*, and a movie titled *"The Imitation Game"*.

P.P. And he is famous because he found the key to crack the German war codes. Turing was one of the inventors of theoretical computer science. At the end, he sadly committed suicide.

Paulo. Tell me what has been going on with the help of computers.

P.P. A hybrid species, the computer mathematicians saw gold in the search of Mersenne primes and this has generated an irrepressible activity.

Take this table home. It is a complete list of all known Mersenne primes — exactly 44 are official. As of June 2015, the last four entries have provisional ranking; not all candidates between M32,582,657 and M57,885,161 have been eliminated.

The last entry in the table is the 48th known Mersenne prime. The largest prime known today is the Mersenne number M_q, with $q = 43112609$; it has 12 978 189 digits.

Table of Mersenne primes M_q

q	Year	Discoverer
2	–	–
3	–	–
5	–	–
7	–	–
13	1456	Anonymous
17	1588	P. A. Cataldi
19	1588	P. A. Cataldi
31	1772	L. Euler
61	1883	I. M. Pervushin
89	1911	R. E. Powers
107	1914	E. Fauquembergue
127	1876	E. Lucas
521	1952	R. M. Robinson
607	1952	R. M. Robinson
1279	1952	R. M. Robinson
2203	1952	R. M. Robinson
2281	1952	R. M. Robinson
3217	1957	H. Riesel
4253	1961	A. Hurwitz
4423	1961	A. Hurwitz
9689	1963	D. B. Gillies
9941	1963	D. B. Gillies
11213	1963	D. B. Gillies
19937	1971	B. Tuckerman
21701	1978	L. C. Noll and L. Nickel
23209	1979	L. C. Noll
44497	1979	H. Nelson and D. Slowinski
86243	1982	D. Slowinski
110503	1988	W. N. Colquitt and L. Welsh, Jr.
132049	1983	D. Slowinski
216091	1985	D. Slowinski
756839	1992	D. Slowinski and P. Gage
859433	1994	D. Slowinski and P. Gage
1257787	1996	D. Slowinski and P. Gage
1398269	1996	J. Armengaud, (G. F. Woltman and GIMPS)
2976221	1997	G. Spence, (G. F. Woltman and GIMPS)
3021377	1998	R. Clarkson, (G. F. Woltman, S. Kurowski and GIMPS)
6972593	1999	N. Hajratwala, (G. F. Woltman, S. Kurowski and GIMPS)
13466917	2001	M. Cameron, (G. F. Woltman, S. Kurowski and GIMPS)

(Continued)

(*Continued*)

q	Year	Discoverer
20996011	2003	M. Shafer, (G. F. Woltman, S. Kurowski and GIMPS)
24036583	2004	J. Findley, (G. F. Woltman, S. Kurowski and GIMPS)
25964951	2005	M. Novak, (G. F. Woltman, S. Kurowski and GIMPS)
30402457	2005	C. Cooper, S. Boone, (G. F. Woltman, S. Kurowski and GIMPS)
32582657	2006	C. Cooper, S. Boone, (G. F. Woltman, S. Kurowski and GIMPS)
37156667	2008	H.-M. Elvenich, (G. F. Woltman, S. Kurowski and GIMPS)
42643801	2009	O. M. Strindmo, (G. F. Woltman, S. Kurowski and GIMPS)
43112609	2008	E. Smith, (G. F. Woltman, S. Kurowski and GIMPS)
57885161	2013	C. Cooper, (G. F. Woltman, S. Kurowski and GIMPS)

In the not-so-distant past, gold and diamond prospectors sacrificed family and friends to go to inhospitable places; jungles with snakes, disease-infested marshes, or high mountains with cliffs and snow, all in search of the precious discovery that would make them rich. The modern searchers of Mersenne primes live a transposed but similar adventure. The location of their findings cannot be anticipated; lucky is the one who first finds IT. No riches, but fame.

Paulo. Papa Paulo, you are a poet.

What do you mean when you said that the location of the findings is not known? It has to be unknown, otherwise everybody would find Mersenne primes.

P.P. Prospecting Mersenne primes has many problems:

(1) It is not known if there are infinitely many Mersenne primes. Since Mersenne primes are being found, it is hard to imagine that suddenly there would be no more. So the belief — I should say the faith — is on infinitely many Mersenne primes.
(2) What is more likely is that there are infinitely many composite Mersenne numbers, which is another difficult question to tackle. Remember, Eric, this was related to your Question 2.

(3) I believe, so I assume, that $M_{q_1} = M_2$, $M_{q_2} = M_3$, $M_{q_3} = M_5, \ldots, M_{q_n}, \ldots$ is the infinite sequence of Mersenne primes written in increasing order.

Given $n \geq 45$, no one has the faintest idea how to find explicitly two positive integers N and k such that M_{q_n} is between N and $N + k$. This explains what I said: The region where large Mersenne primes exist cannot be guessed.

Eric. Total ignorance. Why don't you brains think about the approximate location of Mersenne primes before asking computer people to search? Otherwise it is a blind search.

In an apparent change of subject, Papa Paulo asked:

P.P. I give you a multiple choice question. What is "GIMPS"?

(a) an endangered species of New Zealand wood bears;
(b) celestial bodies which are conglomerates of comets' dust;
(c) an acronym;
(d) particles of subnuclear dimension, whose existence is anticipated by the newest theory of the lunatic physicists.

After some reflection Eric said:

Eric. The answer is (c), even though I don't know what "acronym" means. Knowing how you like to tease, Papa Paulo, it was definitely not (b) or invented only to confuse, (d). But it could be (a). The problem is that I don't know much about endangered species — so (a) had to be excluded.

Paulo. All species are now in danger. Someday only men and viruses will remain on Earth.

Papa Paulo addressed Eric:

P.P. You are right, Eric. "Acronym" means a word formed with the first letters of an expression. In this case, GIMPS is the acronym of "Great Internet Mersenne Primes Search."

Eric. What an idea! How does it work?

P.P. The aim of the GIMPS is to discover large Mersenne primes. Any willing person may participate. They receive the software and a territory to explore, that is, a range of primes q. The participant searches with his personal computer a Mersenne prime M_q, where q is in the given territory. Presently the project has recruited several thousand participants throughout the world. The software used is a specially conceived program to efficiently deal with the large numbers involved. The technical aspects are very sophisticated and involve the unavoidable FFT.

Paulo. FFT is an acronym for ... ?

P.P. Fast Fourier Transform, a trick to perform multiplication of very large numbers.

Eric. Caramba! Long dead, but still giving big trouble.

Paulo. It is absolutely incredible that there are people, not a few people, who spend their time searching for new Mersenne primes. Just to hold the record until it is supplanted by another. But in the meantime to enjoy a moment of glory.

P.P. These people are moved by vanity, a certain thirst for discovery, an addiction like gambling.

Paulo. It is not different from the efforts to surpass sport records.

P.P. The matter of records is not my personal preference, but I give an accolade to all these people. As they search Mersenne primes, they perform very complex calculations with large numbers, we all acquire a better understanding about numbers. Only when theory and computation help each other, working together, substantial progress can be achieved.

Notes About Marin Mersenne (1588–1648)

Mersenne was a member of the religious Ordre des Minimes and spent most of his life in Paris. He published important treatises on the philosophy of science and the acoustical foundations of music. Mersenne maintained correspondence with many illustrious philosophers and scientists in various European countries, as well as in France. For a long time, Mersenne was a favorite correspondent of Fermat and he transmitted problems proposed by Fermat to various mathematicians.

17

Paying My Debt

P.P. I had promised to prove the primality criterion for Mersenne numbers; much preparation will be needed. I will begin by introducing the type of numbers which will be needed in the sequel. They are not rational numbers.

In his teaching, Pythagoras asserted that every segment had measure expressed by a rational number. However, one of his disciples argued as follows: He considered the right triangle with legs measuring 1 and the hypotenuse measuring h. So $h^2 = 1^2 + 1^2 = 2$. If h is a rational number, $h = \frac{m}{n}$, with m, n positive coprime integers, then $\frac{m^2}{n^2} = h^2 = 2$, so $m^2 = 2n^2$. By Euclid's factorization theorem $m = 2^e m'$, where m' is odd and $e \geq 0$; similarly, $n = 2^f n'$, where n' is odd and $f \geq 0$. Thus $2^{2e+1} m'^2 = 2^{2f} n'^2$. By the uniqueness of factorization, $2e + 1 = 2f$, which is impossible. Thus, h is not a rational number. We say that h is an irrational number. Since $h^2 = 2$ we say that h is a square root of 2 and we write $h = \sqrt{2}$. In this same way we may consider the irrational number $\sqrt{3}$, and many more. For every positive integer n, which is not a square, we may consider numbers of the form $r + s\sqrt{n}$, where r and s are rational numbers.

We show with numerical examples how to calculate with these numbers.

$$(3 + 2\sqrt{2}) + (1 + 5\sqrt{2}) = 4 + 7\sqrt{2},$$
$$(3 + 2\sqrt{2}) - (1 + 5\sqrt{2}) = 2 + (-3)\sqrt{2}.$$

This last one I write $2 - 3\sqrt{5}$.

$(3 + 2\sqrt{2}) \times (1 + 5\sqrt{2}) = 3 + 15\sqrt{2} + 2\sqrt{2} + 10(\sqrt{2})^2 = 23 + 17\sqrt{2}$ because $(\sqrt{2})^2 = 2$.

Division is also possible: $3 + 2\sqrt{2}$ divided by $1 + 5\sqrt{2}$ is equal to

$$\frac{3 + 2\sqrt{2}}{1 + 5\sqrt{2}} = \frac{(3 + 2\sqrt{2}) \times (1 - 5\sqrt{2})}{(1 + 5\sqrt{2}) \times (1 - 5\sqrt{2})}.$$

But $(1 + 5\sqrt{2}) \times (1 - 5\sqrt{2}) = 1 - 50 = -49$ so the above expression is equal to $\frac{-17 - 13\sqrt{2}}{-49} = \frac{17}{49} + \frac{13}{49}\sqrt{2}$.

Paulo. Easy. Do this as if the numbers $r + s\sqrt{2}$ were rational numbers when doing operations, remembering that $(\sqrt{2})^2 = 2$.

Eric. I suppose you can do the same with $\sqrt{3}$, $\sqrt{5}$, $\sqrt{6}$ and so on.

P.P. You can. All numbers $r + s\sqrt{d}$ are possible. Here d is any positive integer, which has no square factor except 1.

Eric. And I guess that there are other kinds of numbers?

P.P. You are right, but I shall not discuss these numbers, they are not part of my debt.

Paulo. More numbers? All this to pay a debt about prime numbers. Astonishing.

The Newton Binomial Formula

P.P. This formula gives the powers $(a+b)^n$ where $n > 1$. For example:

$(a + b)^2 = a^2 + 2ab + b^2$
$(a + b)^3 = (a + b)(a^2 + 2ab + b^2) = a^3 + 3a^2b + 3ab^2 + b^3$
$(a + b)^4 = (a + b)(a^3 + 3a^2b + 3ab^2 + b^3)$
$ = a^4 + 4a^3b + 6a^2b^2 + 4ab^3 + b^4.$

Eric. I see what is going on. If $n \geq 2$ then $(a + b)^n$ will be $a^n +$ (a multiple of $a^{n-1}b$) + (a multiple of $a^{n-2}b^2$) + ... you go on until you add b^n. The intermediate terms are multiples of $a^e b^f$ where e, f are positive and $e + f = n$. The only problem is to find the multiples of $a^{n-1}b, a^{n-2}b^2, \ldots, a^e b^f, \ldots$.

P.P. That is indeed the task to be done. Let me first give some notation. The multiplicator of each product $a^e b^f$ is a positive integer. It has to do with n and with the exponent f of b. I may write C_f^n as people did in the past. Today it is more common to write $\binom{n}{f}$. These numbers are called the *binomial coefficients*. Anyway, we can write

$$(a+b)^n = a^n + \binom{n}{1}a^{n-1}b + \binom{n}{2}a^{n-2}b^2 + \cdots + \binom{n}{f}a^e b^f$$

$$+ \cdots + \binom{n}{n-2}a^2 b^{n-2} + \binom{n}{n-1}ab^{n-1} + b^n.$$

We have to find $\binom{n}{1}, \binom{n}{2}, \ldots, \binom{n}{f}, \ldots, \binom{n}{n-2}, \binom{n}{n-1}$.

Paulo. We already did for $n = 2, 3$ and 4

$$\binom{2}{1} = 2$$

$$\binom{3}{1} = 3 \quad \binom{3}{2} = 3$$

$$\binom{4}{1} = 4 \quad \binom{4}{2} = 6 \quad \binom{4}{3} = 4.$$

P.P. To discover $\binom{n}{f}$ (where $2 \leq f \leq n-1$), we see how we get $a^e b^f$ in the power $(a+b)^n$. It comes from $a \times \binom{n-1}{f}a^{e-1}b^f$ and from $b \times \binom{n-1}{f-1}a^e b^{f-1}$. Therefore $\binom{n}{f} = \binom{n-1}{f} + \binom{n-1}{f-1}$.

Please check for $n = 4$. $f = 2, 3$ to see if I am right.

Paulo. Yes, $\binom{4}{2} = \binom{3}{2} + \binom{3}{1}$, $\binom{4}{3} = \binom{3}{3} + \binom{3}{2}$. What is next?

P.P. Now I try to guess what the formula could be for $\binom{n}{f}$. It has to hold for $n = 2, 3, 4$ at least.

Eric. I observed Paulo's calculation and I see that:

$$\binom{2}{1} = 2 = \frac{2}{1}$$

$$\binom{3}{1} = 3 = \frac{3}{1}$$

$$\binom{3}{2} = 3 = \frac{3 \times 2}{1 \times 2}$$

$$\binom{4}{1} = 4 = \frac{4}{1}$$

$$\binom{4}{2} = 6 = \frac{4 \times 3}{1 \times 2}$$

$$\binom{4}{3} = 4 = \frac{4 \times 3 \times 2}{1 \times 2 \times 3}.$$

I guess the formula is

$$\binom{n}{f} = \frac{\text{product of } f \text{ numbers } n, n-1, \ldots, n-(f-1)}{f!}.$$

P.P. You are right. It is easier to write $\binom{n}{f} = \frac{n!}{f!(n-f)!}$. It is just the same, as you observe $\frac{n!}{(n-f)!} = n(n-1)\ldots(n-(f-1))$.

Proof: Suppose we already know that the formula holds for $2, 3, \ldots, n-1$ and all f, $1 \le f \le n-1$, respectively. It is OK because it holds for $n = 2$, $f = 1$. Then,

$$\binom{n}{f} = \binom{n-1}{f} + \binom{n-1}{f-1} = \frac{(n-1)!}{f!(n-1-f)!} + \frac{(n-1)!}{(f-1)!(n-f)!}.$$

The least common multiple of the denominators is
$$f!(n-f)! = f!(n-1-f)! \times (n-f), \text{ but also,}$$
$$f!(n-f)! = f \times (f-1)!(n-f)!.$$

Hence, the sum to calculate is

$$\frac{(n-1)!(n-f)}{f!(n-f)!} + \frac{(n-f)!f}{f!(n-f)!} = \frac{(n-1)!n}{f!(n-f)!} = \frac{n!}{f!(n-f)!}.$$

So we proved that the formula for $\binom{n}{f}$ is still true. By the method of induction, the formula can never be false. It is always true.

q.e.d.

Paulo. That was very neat. Who found the formula?

P.P. It is attributed to Newton.

There is a famous triangle — not Pythagoras' triangle — made up with these binomial coefficients. It is a special configuration with the binomial coefficients $\binom{n}{f}$, which Pascal put forth:

$$\begin{array}{ccccccccccccc}
 & & & & & & 1 & & & & & & \\
 & & & & & 1 & & 1 & & & & & \\
 & & & & 1 & & 2 & & 1 & & & & \\
 & & & 1 & & 3 & & 3 & & 1 & & & \\
 & & 1 & & 4 & & 6 & & 4 & & 1 & & \\
 & 1 & & 5 & & 10 & & 10 & & 5 & & 1 & \\
1 & & 6 & & 15 & & 20 & & 15 & & 6 & & 1
\end{array}$$

etc.

For example, the numbers in row 6 are $\binom{6}{0} = 1$, $\binom{6}{1} = 6$, $\binom{6}{2} = 15$, $\binom{6}{3} = 20$, $\binom{6}{4} = 15$, $\binom{6}{5} = 6$, $\binom{6}{6} = 1$.

Paulo. What do you do with *Pascal's triangle*?

P.P. You play. There are many relations between these numbers — but I'll give only a few facts, which are obvious from the formula

$$\binom{n}{f} = \binom{n}{n-f}$$

when $1 \leq f \leq n-1$. By convention we like to start with $\binom{n}{0} = 1$ when $n \geq 1$. The purpose is simple. We may write Newton's binomial formula as a summation $(a+b)^n = \sum_{f=0}^{n} \binom{n}{f} a^{n-f} b^f$.

If we take $a = 1$ and $b = 1$, then

$$2^n = 1 + \binom{n}{1} + \binom{n}{2} + \cdots + \binom{n}{n-1} + 1.$$

This is the sum of terms in the row n of Pascal's triangle. If we take $a = 1$ and $b = -1$, then

$$0 = 1 - \binom{n}{1} + \binom{n}{2} - \binom{n}{3} + \cdots + (-1)^{n-1}\binom{n}{n-1} + (-1)^n.$$

Paulo. It's good to play with Newton's binomial formula. I see you can get many more relations between the binomial coefficients, choosing for example $a = 2$, $b = 1$, or $a = 2$, $b = -1$, etc.

P.P. Now I will give a divisibility property of binomial coefficients. It is as easy to prove as it is important. Let p be a prime, let k be such that $1 \leq k \leq p - 1$. Then p divides $\binom{p}{k}$.

Eric. I see why. You use the formula and think just a minute...

$$k\binom{p}{k} = k \times \frac{p(p1)\ldots(p-k+1)}{1 \times 2 \times \cdots \times k}$$

$$= k \times \frac{p}{k} \times \frac{(p-1)(p-2)\ldots(p-k+1)}{1 \times 2 \times \cdots \times (k-1)} = p\binom{p-1}{k-1}.$$

But $\binom{p-1}{k-1}$ is an integer, so p divides $k\binom{p}{k}$ — and $1 \leq k \leq p-1$, so p does not divide k. Therefore p divides $\binom{p}{k}$. q.e.d.

Paulo. The proof is good, but Papa Paulo, you were bragging and trying to impress us. That took more than one minute.

P.P. Yes. More than one minute to write, but less than 15 seconds to think.

What a braggart Papa Paulo is!
Papa Paulo continued.

P.P. Now take home these notes about Pythagoras and Pascal. They describe some of their mathematical discoveries along with other interesting facts.

Recurring Sequences

P.P. Remember, the primality criterion for Mersenne numbers required a comparison with the numbers $S_0 = 4$, $S_1 = S_0^2 - 2 = 14$, $S_2 = S_1^2 - 2 = 194$, etc.

These numbers are terms of a recurring sequence. So it is necessary, as part of my debt, to explain what these recurring sequences are.

Once, I considered numbers obtained by jumping by 8, like 3 11 19 27 35.... Each term of the sequence is perfectly determined and easy to calculate. For example, the 100th term is just $3 + 99 \times 8 = 795$ in the same way, given a number a (like $a = 3$) and a number $d \geq 1$ (like $d = 8$), it is possible to consider the sequence a, $a + d$, $a + 2d$, $a + 3d, \ldots$. This is called an *arithmetic progression*, with *initial term* a and *difference* d. Arithmetic progressions are very important in connection with prime numbers, as I will say later in our discussions. But they are too simple and not useful for the primality criterion for Mersenne numbers.

We shall need binary recurring sequences. The original example is the sequence of *Fibonacci numbers*. I must use a notation which might be confusing, but it is traditional. The Fibonacci numbers appeared in the solution of a problem considered by Fibonacci in his influential book *"Liber Abaci"*, which was published in 1202.

Eric. The title of that book sounds Latin. What does it mean?

P.P. Just *"Book of Calculations"*.

Paulo. I have never met anyone who reads Latin. How can one read this book today?

P.P. It was a very important book, so it has been translated into most modern languages. You can find many problems proposed and discussed in full, including the problem of rabbits. Fibonacci numbers are not relevant to paying my debt. For that, I will need other sequences. Nevertheless, the Fibonacci numbers are famous, so I thought you would enjoy learning a little bit about them.

Eric. Now, please tell us the rabbits problem.

P.P. The story is about rabbits in an enclosed space. On January 1st there is one male and one female rabbit in the enclosure. The couple has no offspring in January or February, but produces a new couple (male, female) in March, and each month thereafter. Each new couple matures in their first two months, then produces a new couple in their third month, and each month thereafter. It is assumed that no rabbit dies before the end of the year. The question is: How many couples of rabbits there are at the end of the year?

Eric. It is an easy problem to solve, just keep track of the number of births. Let me get organized. The best is that I give a name or a mark to each couple of rabbits, like $A, B, C, \ldots$. I also have to take into account the births. I will mark $B \leftarrow A$ when the couple B is an offspring of the couple A. Check my table:

Month	Couples	Number
January	A	1
February	A	1
March	$A\ B \leftarrow A$	2
April	$A\ B\ C \leftarrow A$	3
May	$A\ B\ C\ D \leftarrow A$	
	$E \leftarrow B$	5
June	$A\ B\ C\ D\ E$	
	$F \leftarrow A,\ G \leftarrow B,\ H \leftarrow C$	8

Eric (*continued*). Now I see how the numbers increase; each number is the sum of the numbers from the two preceding months. Reason, say for July, you have to add the number of those in June (old and new couples) plus the number of old couples in June, which is the number of couples in May, because they can procreate. So in July there are 13 couples. The sequence of numbers of couples is

$$1\ \ 1\ \ 2\ \ 3\ \ 5\ \ 8\ \ 13\ \ 21\ \ 34\ \ 55\ \ 89\ \text{ and } 144,$$

which is the number of couples of rabbits in December.

P.P. Eric, you just stated the Fibonacci numbers, you put $F_0 = 0$, $F_1 = 1,\ F_2 = 1, \ldots.\ F_n = F_{n-1} + F_{n-2}$ for all $n \geq 2$. This definition is good to determine all the Fibonacci numbers.

Eric. But it is quite long, if I want to find the number F_{100}. I know you mathematicians must have some formula to compute F_{100} or F_n for any integer n, however large n may be.

P.P. Yes we do, because the terms of the sequence are obtained with a simple and repetitive law, $F_n = F_{n-1} + F_{n-2}$. This is completely different from the situation for the sequence of prime numbers.

Eric. Is the formula for F_n difficult? Can you explain it?

P.P. I think I have to. I don't want to be in debt when I later use the Fibonacci numbers. The formula for Fibonacci numbers is expressed in terms of two surds, $\gamma = \frac{1+\sqrt{5}}{2}$ and $\delta = \frac{1-\sqrt{5}}{2}$.

The number γ is called the *golden ratio*, and it has a most interesting history. The golden ratio appeared in questions of geometry at the time of the Greeks. It is the central character of many books, which I leave for you to discover. That discussion is not part of my debt, even though I need the golden ratio as part of my payment. I will give the formula for F_n which is in terms of γ and δ.

Eric. It is curious how the two roots γ and δ, which are not natural numbers, produce natural numbers.

Paulo. No mystery, it all depends how they are put together, like $\gamma + \delta = \frac{1+\sqrt{5}}{2} + \frac{1-\sqrt{5}}{2} = 1$, but it is not good for $\gamma - \delta = \frac{1+\sqrt{5}}{2} - \frac{1-\sqrt{5}}{2} = \sqrt{5}$, and again good for

$$\gamma\delta = \frac{1+\sqrt{5}}{2} \times \frac{1-\sqrt{5}}{2} = -1.$$

P.P. The Fibonacci numbers constitute a *binary recurring sequence*, because each term is obtained in a prescribed way from the two predecessors. Another binary recurring sequence, intimately connected with the Fibonacci sequence, is formed by the *Lucas numbers* $L_0 = 2$, $L_1 = 1$, $L_2 = 3$, $L_3 = 4, \ldots$ and $L_n = L_{n-1} + L_{n-2}$ for all $n \geq 2$. The sequences of Fibonacci and of Lucas numbers have the same law of recurrence, but different initial terms.

Now I shall give the *formulas for F_n and L_n*. First I observe that $\gamma^2 = \gamma + 1$ and $\delta^2 = \delta + 1$. This is easy to check. Paulo, you do it.

Paulo. Sure. I just compute

$$\gamma^2 = \left(\frac{1+\sqrt{5}}{2}\right)^2 = \frac{1}{4}(1+5+2\sqrt{5}) = \frac{3+\sqrt{5}}{2}$$

and

$$\gamma + 1 = \frac{1+\sqrt{5}}{2} + 1 = \frac{1+\sqrt{5}+2}{2} = \frac{3+\sqrt{5}}{2}, \text{ so } \gamma^2 = \gamma + 1.$$

I am sure that the same calculation gives $\delta^2 = \delta + 1$.

P.P. So $\gamma^2 - \gamma - 1 = 0$ and $\delta^2 - \delta - 1 = 0$. This means that the quadratic equation $X^2 - X - 1$ has the solutions γ and δ.

Eric. I could have found it by using the formula to find the solutions of quadratic equations. We all learn this in school. The solutions are

$$\frac{1+\sqrt{1+4}}{2} = \frac{1+\sqrt{5}}{2} = \gamma \quad \text{and} \quad \frac{1-\sqrt{1+4}}{2} = \frac{1-\sqrt{5}}{2} = \delta.$$

P.P. From $\gamma^2 = \gamma + 1$, multiplying by $\gamma, \gamma^2, \gamma^3, \ldots$ we obtain

$$\gamma^3 = \gamma^2 + \gamma, \qquad \gamma^4 = \gamma^3 + \gamma^2, \ldots$$

and for any $n \geq 2$, $\gamma^n = \gamma^{n-1} + \gamma^{n-2}$. In the same way $\delta^n = \delta^{n-1} + \delta^{n-2}$ for $n \geq 2$.

After these preparations, let

$$F'_0 = \frac{\gamma^0 - \delta^0}{\sqrt{5}} = \frac{1-1}{\sqrt{5}} = 0,$$

$$F'_1 = \frac{\gamma - \delta}{\sqrt{5}} = \frac{\sqrt{5}}{\sqrt{5}} = 1,$$

$$F'_2 = \frac{\gamma^2 - \delta^2}{\sqrt{5}} = \frac{(\gamma+1) - (\delta+1)}{\sqrt{5}}$$

$$= \frac{\gamma - \delta}{\sqrt{5}} + \frac{1-1}{\sqrt{5}} = F'_1 + F'_0,$$

$$F'_3 = \frac{\gamma^3 - \delta^3}{\sqrt{5}} = \frac{\gamma^2 - \delta^2}{\sqrt{5}} + \frac{\gamma - \delta}{\sqrt{5}} = F'_2 + F'_1.$$

In the same way, $F'_n = \frac{\gamma^n - \delta^n}{\sqrt{5}} = \frac{\gamma^{n-1} - \delta^{m-1}}{\sqrt{5}} + \frac{\gamma^{n-2} - \delta^{n-2}}{\sqrt{5}} = F'_{n-1} + F'_{n-2}$. I observe that the numbers F'_n form a binary recurring sequence with the same law of recurrence and the same initial terms as the sequence of Fibonacci numbers.

So the two sequences coincide:

$$F_0 = 0 = F'_0$$
$$F_1 = 1 = F'_1$$
$$F_2 = F_1 + F_0 = F'_1 + F'_0 = F'_2$$

and for every $n \geq 2$

$$F_n = F_{n-1} + F_{n-2} = F'_{n-1} + F'_{n-2} = F'_n$$

that is

$$F_n = \frac{\gamma^n - \delta^n}{\sqrt{5}}.$$

Eric. Without telling us, you got the formula for Fibonacci numbers. And what is the formula for Lucas numbers?

P.P. It is $L_n = \gamma^n + \delta^n$ for all $n \geq 0$. I know that you'll be able to prove this formula; the proof is like for F_n.

Eric. But, I am still a bit annoyed with the calculation of the powers of γ and of δ. You told us about a shortcut to calculate powers of natural numbers, but γ and δ are not natural numbers. Any shortcuts to calculate F_n and L_n?

P.P. Of course, we are never short of shortcuts. This time we use 2 by 2 matrices.

Paulo. "Matrices", you never mentioned this word. 2 by 2? What is this?

P.P. Easy. A 2 by 2 matrix is an array of numbers, letters (whatever you want). There are two rows of two things, one below the other,

like
$$\begin{pmatrix} 3 & -1 \\ 0 & 5 \end{pmatrix}$$

So two rows, two columns. If I use letters, I could write
$$\begin{pmatrix} a_{11} & a_{12} \\ a_{21} & a_{22} \end{pmatrix}$$

a_{11} entry in row 1, column 1

a_{12} ...

Paulo interrupted:

Paulo. a_{21} entry in row 2 and column 1 and so on. What will we do with matrices?

P.P. We multiply them. I have to tell you the rule of multiplication. This is better explained with letters. I wish to multiply

$$A = \begin{pmatrix} a_{11} & a_{12} \\ a_{21} & a_{22} \end{pmatrix} \quad \text{with} \quad B = \begin{pmatrix} b_{11} & b_{12} \\ b_{21} & b_{22} \end{pmatrix}$$

to get AB, which is going to be a 2 by 2 matrix, let

$$AB = C = \begin{pmatrix} c_{11} & c_{12} \\ c_{21} & c_{22} \end{pmatrix}.$$

To define the multiplication I have to say how to obtain c_{11}, c_{12}, c_{21} and c_{22}.

This is the rule:
$$c_{11} = a_{11}b_{11} + a_{12}b_{21}$$
$$c_{12} = a_{11}b_{12} + a_{12}b_{22}$$
$$c_{21} = a_{21}b_{11} + a_{22}b_{21}$$
$$c_{22} = a_{21}b_{12} + a_{22}b_{22}.$$

Eric. That is hard to remember. Why this strange rule?

P.P. First, why this rule? It is explained in a natural way in algebra — but this is not part of my debt. Check the books. There are hundreds of books, so spare me from giving titles.

To remember

c_{11}: with a left-hand finger follow the row 1 of A, with a right-hand finger follow the column 1 of B, multiplying corresponding entries and adding the results.

For c_{12}: same, with row 1 of A and column 2 of B.

etc.

Eric. How about people with no hands, or with hands in pockets? Let us do numerical examples.

Paulo. OK, let's do it!

P.P. We multiply $A = \begin{pmatrix} 3 & -1 \\ 0 & 5 \end{pmatrix}$ with $B = \begin{pmatrix} -1 & -1 \\ 3 & 0 \end{pmatrix}$. We get $AB = \begin{pmatrix} -6 & -3 \\ 15 & 0 \end{pmatrix}$ because

$$c_{11} = 3 \times (-1) + (-1) \times 3 = -6,$$
$$c_{12} = 3 \times (-1) + (-1) \times 0 = -3,$$
$$c_{21} = 0 \times (-1) + 5 \times 3 = 15,$$
$$c_{22} = 0 \times (-1) \times 5 \times 0 = 0.$$

You may practice at home. Let

$$A = \begin{pmatrix} -2 & -1 \\ 1 & 2 \end{pmatrix}, \quad B = \begin{pmatrix} 0 & 1 \\ 8 & 3 \end{pmatrix}$$

and also

$$A = \begin{pmatrix} -1 & -3 \\ 4 & 5 \end{pmatrix}, \quad B = \begin{pmatrix} -3 & 4 \\ -1 & 9 \end{pmatrix}.$$

Calculate AB and if you like, calculate also BA and you'll see, what you have never thought possible $AB \neq BA$.

With 2 by 2 matrices, I am interested in calculating powers; of course

$$A^2 = AA, \quad A^3 = A^2 A = AA^2, \quad \text{etc.}$$

Ready to follow the shortcut to calculate F_n and L_n?

Paulo. Yes, but it would be better to do a numerical example. How about F_{21} and L_{21}?

P.P. First write the dyadic expression of 21:

$$21 = 1 \times 2^4 + 1 \times 2^2 + 1.$$

Let

$$A = \begin{pmatrix} 0 & 1 \\ 1 & 1 \end{pmatrix} = \begin{pmatrix} F_0 & F_1 \\ F_1 & F_2 \end{pmatrix}$$

$$A^2 = \begin{pmatrix} 0 & 1 \\ 1 & 1 \end{pmatrix}\begin{pmatrix} 0 & 1 \\ 1 & 1 \end{pmatrix} = \begin{pmatrix} 1 & 1 \\ 1 & 2 \end{pmatrix} = \begin{pmatrix} F_1 & F_2 \\ F_2 & F_3 \end{pmatrix}$$

$$A^3 = \begin{pmatrix} 0 & 1 \\ 1 & 1 \end{pmatrix}\begin{pmatrix} 1 & 1 \\ 1 & 2 \end{pmatrix} = \begin{pmatrix} 1 & 2 \\ 2 & 3 \end{pmatrix} = \begin{pmatrix} F_2 & F_3 \\ F_3 & F_4 \end{pmatrix}$$

$$A^4 = \begin{pmatrix} 0 & 1 \\ 1 & 1 \end{pmatrix}\begin{pmatrix} 1 & 2 \\ 2 & 3 \end{pmatrix} = \begin{pmatrix} 2 & 3 \\ 3 & 5 \end{pmatrix} = \begin{pmatrix} F_3 & F_4 \\ F_4 & F_5 \end{pmatrix}.$$

If we already know that $A^{n-1} = \begin{pmatrix} F_{n-2} & F_{n-1} \\ F_{n-1} & F_n \end{pmatrix}$, then

$$A^n = A\, A^{n-1} = \begin{pmatrix} 0 & 1 \\ 1 & 1 \end{pmatrix}\begin{pmatrix} F_{n-2} & F_{n-1} \\ F_{n-1} & F_n \end{pmatrix}$$

$$= \begin{pmatrix} 0 \times F_{n-2} + 1 \times F_{n-1} & 0 \times F_{n-1} + 1 \times F_n \\ 1 \times F_{n-2} + 1 \times F_{n-1} & 1 \times F_{n-1} + 1 \times F_n \end{pmatrix}$$

$$= \begin{pmatrix} F_{n-1} & F_n \\ F_n & F_{n+1} \end{pmatrix}.$$

Now to calculate F_{21}: it is the entry in the first row and second column of $A^{21} = A^{2^4+2^2+1} = A^{2^4} \times A^{2^2} \times A$. We obtain the factors by successive squaring:

$$A = \begin{pmatrix} 0 & 1 \\ 1 & 1 \end{pmatrix} \qquad A^2 = \begin{pmatrix} 1 & 1 \\ 1 & 2 \end{pmatrix}$$

$$A^4 = \begin{pmatrix} 2 & 3 \\ 3 & 5 \end{pmatrix} \qquad A^8 = \begin{pmatrix} 13 & 21 \\ 21 & 34 \end{pmatrix}$$

$$A^{16} = \begin{pmatrix} 610 & 787 \\ 987 & 1597 \end{pmatrix}$$

and finally
$$A^{21} = A^{16} \times A^4 \times A = \begin{pmatrix} 610 & 987 \\ 987 & 1597 \end{pmatrix} \times \begin{pmatrix} 3 & 5 \\ 5 & 8 \end{pmatrix}$$
$$= \begin{pmatrix} 6765 & 10946 \\ 10946 & 17711 \end{pmatrix}$$

so $F_{21} = 10946$.

Eric. It is clear what to do. The numbers soon become big, but computers handle this easily up to the point they can't accumulate the numbers any more because of their sizes.

What is L_{21}?

P.P. You can do it by yourself, just begin with $B = \begin{pmatrix} 2 & 1 \\ 1 & 3 \end{pmatrix} = \begin{pmatrix} L_0 & L_1 \\ L_1 & L_2 \end{pmatrix}$, prove that $BA^n = \begin{pmatrix} L_{n-1} & L_n \\ L_n & L_{n+1} \end{pmatrix}$ and L_{21} is in the first row and second column of $B \times A^{21}$. The result is $L_{21} = 24476$.

Eric. The numbers are pretty big. Can one have some idea of the size of F_n and L_n?

P.P. Yes. I'll just say that the number of digits of F_n or L_n is about $\frac{n}{5}$.

There was a silence, which showed that Paulo and Eric (and me, too) were tired of this deviation. I resumed:

P.P. I hope you appreciate how nice I have been, paying more than was my debt! There are two binary recurring sequences which — you will be surprised — are closely connected to my debt, more precisely, with Fermat numbers and with Mersenne numbers.

Eric. Are you going to tell us that Fermat numbers and Mersenne numbers form binary recurring sequences?

P.P. They do not, but they are numbers appearing in two binary recurring sequences. Listen, let
$P = 3$, $Q = 2$, $U_0 = 0$, $U_1 = 1$, then $U_2 = 3 \times 1 - 2 \times 0 = 3 = 2^2 - 1$, $U_3 = 3 \times 3 - 2 \times 1 = 7 = 2^3 - 1$, $U_4 = 3 \times 7 - 2 \times 3 = 15 = 2^4 - 1$ and it is easy to see that $U_n = 3U_{n-1} - 2U_{n-2} = 2^n - 1$, for every $n \geq 2$.

Indeed, if we already know that $U_{n-1} = 2^{n-1} - 1$ and $U_{n-2} = 2^{n-2} - 1$ then $U_n = 3 \times U_{n-1} - 2U_{n-2} = 3 \times (2^{n-1} - 1) - 2 \times (2^{n-2} - 1) = 3 \times 2^{n-1} - 3 - 2 \times 2^{n-2} + 2 = 2 \times 2^{n-1} - 1 = 2^n - 1$. This sequence includes the Mersenne numbers.

In the same way, if $P = 3$, and $Q = 2$, let $V_0 = 2$, $V_1 = 3$, then $V_2 = 3 \times 3 - 2 \times 2 = 5 = 2^2 + 1$, $V_3 = 3 \times 5 - 2 \times 3 = 9 = 2^3 + 1$ and more generally, $V_n = 2^n + 1$ for all $n \geq 2$.

I'll let you do this proof. So this sequence includes the Fermat numbers.

And now I'll finally start paying my debt. Remember I have to justify with a complete proof the primality criterion for Mersenne numbers. This will require many steps, serving as support of subsequent steps. I will not deviate from the direct line of proof. All the lemmas are very special particular cases of results which are known for sequences considered by Lucas in his important papers, of which the long American paper of 1878 is the most influential.

Eric. Wasn't Lucas a Frenchman? Why did he not publish his long and important paper in France?

P.P. He was looked down on by the establishment. Influential mathematicians dismissed his work as research of fringe interests or recreations. How wrong! This happens even in science; let us not speak about music and painting, when innovators explore totally new avenues. Today, the ideas originated by Lucas occupy a central position. Of course, there are now much more sophisticated methods; such is the progress of science. So, let us honor this man. Are you ready? I will start my payment. I consider two binary recurring sequences:

(1) $U_0 = 0$, $U_1 = 1$, $U_2 = 2$, $U_3 = 6, \ldots, U_n = 2U_{n-1} + 2U_{n-2}$ for all $n \geq 2$, and
(2) $V_0 = 2$, $V_1 = 2$, $V_2 = 8$, $V_3 = 20, \ldots, V_n = 2V_{n-1} + 2V_{n-2}$ for all $n \geq 2$.
(3) Let $\alpha = 1 + \sqrt{3}$ and $\beta = 1 - \sqrt{3}$. Then $\alpha + \beta = 2$, $\alpha\beta = (1 + \sqrt{5}) \times (1 - \sqrt{3}) = 1 - 3 = -2$, $\alpha - \beta = 2\sqrt{3}$, $\alpha^2 = 2\alpha + 2$,

because $(1+\sqrt{3})^2 = (1+2\sqrt{3}) + 3 = 2(1+\sqrt{3}) + 2 = 2\alpha + 2$.
Similarly $\beta^2 = 2\beta + 2$.

(4) For every $n \geq 0$:
$$U_n = \frac{\alpha^n - \beta^n}{2\sqrt{3}} \quad \text{and} \quad V_n = \alpha^n + \beta^n.$$

Proof: The proof is exactly like the proof of the formulas for the Fibonacci and Lucas numbers. I repeat it without too many explanations.

Let $U'_n = \frac{\alpha^n - \beta^n}{2\sqrt{3}}$, $V'_n = \alpha^n + \beta^n$, then

$U'_0 = 0 = U_0$,

$U'_1 = 1 = U_1$,

$$U'_n = \frac{\alpha^n - \beta^n}{2\sqrt{3}} = 2\frac{\alpha^{n-1} - \beta^{n-1}}{2\sqrt{3}} + 2\frac{\alpha^{m-2} - \beta^{n-2}}{2\sqrt{3}}$$

$$= 2U'_{n-1} + 2U'_{n-2} = 2U_{n-1} + 2U_{n-2} = U_n \quad \text{for all } n \geq 2.$$

The proof of the formula for V_n is similar. q.e.d.

Eric. Papa Paulo, you had a good idea to first show the proof for the Fibonacci and Lucas numbers. This one was almost the same proof, so it was easy to follow. What is next?

P.P. Now I will give relations between the terms of the sequence of U's and also of the sequence of V's. You will note that for the U's I need the V's, for the V's I need the U's. This is why I study both sequences simultaneously.

(5) For every $n \geq 0$:
$$V_n^2 - 12 U_n^2 = (-2)^{n+2}.$$

Proof: The proof uses the formulas for U_n and V_n:

$$V_n^2 - 12 U_n^2 = (\alpha^n + \beta^n)^2 - 12 \times \left(\frac{\alpha^n - \beta^n}{2\sqrt{3}}\right)^2$$

$$= \alpha^{2n} + 2\alpha^n \beta^n + \beta^{2n} - 12 \times \frac{\alpha^{2n} - 2\alpha^n \beta^n + \beta^{2n}}{12}.$$

By (3) $\alpha^n \beta^n = (-2)^n$, hence the preceding expression, after cancellations becomes equal to $4(-2)^n = (-2)^{n+2}$. q.e.d.

One good practice for both of you is to choose values for n then see if the formula is good. Take, for example $n = 5$, then $V_5 = 152$, $U_5 = 44$ and $V_5^2 - 12 \times U_5^2 = 152^2 - 12 \times 44^2 = 23104 - 12 \times 1936 = 23104 - 23232 = -128 = (-2)^7$. It worked.

(6) $U_{2n} = U_n V_n$ for all $n \geq 0$.

Proof: $U_n V_n = \dfrac{\alpha^n - \beta^n}{2\sqrt{3}} \times (\alpha^n + \beta^n) = \dfrac{\alpha^{2n} - \beta^{2n}}{2\sqrt{3}} = U_{2n}$. q.e.d.

(7) $V_{2n} = V_n^2 + (-2)^{n+1}$.

Proof:

$$\begin{aligned} V_n^2 + (-2)^{n+1} &= (\alpha^n + \beta^n)^2 + (-2)^{n+1} \\ &= \alpha^{2n} + \beta^{2n} + 2(\alpha\beta)^n - 2(-2)^n \\ &= \alpha^{2n} + \beta^{2n} = V_{2n}. \end{aligned}$$

q.e.d.

(8) If $m \geq n \geq 0$ then $U_{m+n} = U_m V_n - (-2)^n U_{m-n}$.

Proof:

$$\begin{aligned} &U_m V_n - (-2)^n U_{m-n} \\ &= \frac{1}{2\sqrt{3}}\left((\alpha^m - \beta^m)(\alpha^n + \beta^n)\right) - (-2)^n(\alpha^{m-n} - \beta^{m-n}) \\ &= \frac{1}{2\sqrt{3}}\left[\alpha^{m+n} - \beta^{m+n} + \alpha^m \beta^n - \alpha^n \beta^m - \alpha^n \beta^n(\alpha^{m-n} - \beta^{m-n})\right] \\ &= \frac{1}{2\sqrt{3}}(\alpha^{m+n} - \beta^{m-n}) = U_{m+n}. \end{aligned}$$

q.e.d.

(9) If $m \geq n \geq 0$ then $U_m V_n - U_n V_m = 2 \times (-2)^n U_{m-n}$.

Proof:

$$\begin{aligned} &U_m V_n - U_n V_m \\ &= \frac{1}{2\sqrt{3}}\left[(\alpha^m - \beta^m) \times (\alpha^n + \beta^n) - (\alpha^n - \beta^n) \times (\alpha^m + \beta^m)\right] \end{aligned}$$

$$= \frac{1}{2\sqrt{3}}\left[(\alpha^{m+n} - \beta^{m+n}) + \alpha^m\beta^n - \alpha^n\beta^m - (\alpha^{m+n} - \beta^{m+n})\right.$$
$$\left. - \alpha^n\beta^m + \alpha^m\beta^n\right]$$
$$= \frac{1}{2\sqrt{3}}(2\alpha^m\beta^n - 2\alpha^n\beta^m)$$
$$= \frac{1}{2\sqrt{3}}2(-2)^n(\alpha^{m-n} - \beta^{m-n})$$
$$= 2(-2)^n U_{m-n}. \qquad \text{q.e.d.}$$

(10) If $m, n \geq 0$ then $2V_{m+n} = V_m V_n + 12 U_m U_n$.

Proof:

$$V_m V_n + 12 U_m U_n$$
$$= (\alpha^m + \beta^m)(\alpha^n + \beta^n) + 12 \frac{\alpha^m - \beta^m}{2\sqrt{3}} \times \frac{\alpha^n - \beta^n}{2\sqrt{3}}$$
$$= \alpha^{m+n} + \beta^{m+n} + \alpha^m\beta^n + \alpha^n\beta^m +$$
$$12\frac{\alpha^{m+n} + \beta^{m+n} - \alpha^m\beta^n - \alpha^n\beta^n}{12}$$
$$= 2(\alpha^{m+n} + \beta^{m+n}) = 2V_{m+n}. \qquad \text{q.e.d.}$$

Paulo. Stop, stop! I have never seen so many relations between numbers. Papa Paulo, are you going to need all these relations?

P.P. Yes, sir. These relations are just a few among the myriad of known relations and there are still more to come, more being discovered.

Paulo. OK! But what is the meaning of the word "myriad"? Ten, hundred, thousand?

P.P. Well, I say "myriad" of something when I cannot count how many there are.

Eric, who didn't miss an opportunity:

Eric. Papa Paulo: A mathematician who cannot count!

P.P. (*I was caught and I had to defend myself...*). It is a myriad of relations because it is always possible to combine relations already known to obtain new ones. But ultimately, they all come from the defining recurrences $U_n = 2U_{n-1} + 2U_{n-2}$ and $V_n = 2V_{n-1} + 2V_{n-2}$.

(11) For $n \geq 1$: $U_n = \binom{n}{1} + \binom{n}{3} \times 3 + \binom{n}{5} \times 3^2 + \binom{n}{7} \times 3^3 + \cdots +$ last summand, where the last summand

$$= \begin{cases} \binom{n}{n-1} \times 3^{\frac{n-2}{2}} & \text{if } n \text{ is even} \\ 3^{\frac{n-1}{2}} & \text{if } n \text{ is odd} \end{cases}$$

Proof: $2\sqrt{3} \times U_n = \alpha^n - \beta^n = (1+\sqrt{3})^n - (1-\sqrt{3})^n$. Now I use the expression for the nth power of a sum or difference:

$$(1+\sqrt{3})^n = 1 + \binom{n}{1}\sqrt{3} + \binom{n}{2}\sqrt{3}^2 + \binom{n}{3}\sqrt{3}^3$$

$$+ \cdots + \binom{n}{n-1}\sqrt{3}^{n-1} + \sqrt{3}^n$$

and

$$(1-\sqrt{3})^n = 1 - \binom{n}{1}\sqrt{3} + \binom{n}{2}\sqrt{3}^2 - \binom{n}{3}\sqrt{3}^3$$

$$+ \cdots + (-1)^{n-1}\binom{n}{n-1}\sqrt{3}^{n-1} + (-1)^n \sqrt{3}^n.$$

By subtraction, and division by $2\sqrt{3}$ we get $U_n = \binom{n}{1} + \binom{n}{3} \times 3 + \binom{n}{5} \times 3^2 + \cdots +$ last summand, where the last summand is the one indicated. q.e.d.

(12) For $m \geq 1$ and q odd:

$$12^{\frac{q-1}{2}} U_m^q = U_{mq} - \binom{q}{1}(-2)^m U_{m(q-2)} + \binom{q}{2}(-2)^{2m} U_{m(q-4)}$$

$$- \cdots + (-1)^{\frac{q-1}{2}} \binom{q}{\frac{q-1}{2}}(-2)^{\frac{q-1}{2} \times m} U_m.$$

Proof:
$$12^{\frac{q-1}{2}} U_m^q = (2\sqrt{3})^{q-1} \left(\frac{\alpha^m - \beta^m}{2\sqrt{3}}\right)^q$$

$$= \frac{1}{2\sqrt{3}}\Big[\alpha^{mq} - \binom{q}{1}\alpha^{m(q-1)}\beta^m + \binom{q}{2}\alpha^{m(q-2)}\beta^{2m}$$

$$- \cdots + \binom{q}{2}\alpha^{2m}\beta^{2m}\left(\alpha^{m(q-4)} - \beta^{m(q-4)}\right)\cdots$$

$$+ (-1)^{\frac{q-1}{2}}\binom{q}{\frac{q-1}{2}}\alpha^{m\frac{q-1}{2}}\beta^{m\frac{q-1}{2}}(\alpha^m - \beta^m)\Big].$$

Hence the above expression is equal to

$$U_{mq} - \binom{q}{1}(-2)^m U_{m(q-2)} + \binom{q}{2}(-2)^{2m} U_{m(q-4)} - \cdots$$

$$U_n V_n = \frac{\alpha^n - \beta^n}{2\sqrt{3}} \times (\alpha^n + \beta^n) = \frac{\alpha^{2n}}{2\sqrt{3}}$$

$$= U_{2n} + (-1)^{\frac{q-1}{2}}\binom{q}{\frac{q-1}{2}}(-2)^{\frac{q-1}{2}} U_m. \qquad\text{q.e.d.}$$

(13) For q odd:

$$2^q = V_q - \binom{q}{1} \times 2V_{q-2} + \binom{q}{2} \times 2^2 V_{q-4} - \binom{q}{3} \times 2^3 V_{q-6} \cdots$$

$$+ (-1)^{\frac{q-1}{2}}\binom{q}{\frac{q-1}{2}} \times 2^{\frac{q-1}{2}} V_1.$$

Proof:

$$2^q = (\alpha + \beta)^q = \alpha^q + \binom{q}{1}\alpha^{q-1}\beta + \binom{q}{2}\alpha^{q-2}\beta^2 + \cdots$$

$$+ \binom{q}{q-2}\alpha^2\beta^{q-2} + \binom{q}{q-1}\alpha\beta^{q-1} + \beta^q.$$

Grouping summands two by two like in the preceding proof, we get

$$2^q = (\alpha^q + \beta^q) + \binom{q}{1}\alpha\beta(\alpha^{q-2} + \beta^{q-2}) + \binom{q}{2}(\alpha\beta)^2(\alpha^{q-4} + \beta^{q-4})$$

$$+ \cdots + \binom{q}{\frac{q-1}{2}}(\alpha\beta)^{\frac{q-1}{2}}(\alpha + \beta)$$

$$V_q = \binom{q}{1} \times 2V_{q-2} + \binom{q}{2} \times 2^2 V_{q-4}$$
$$+ \cdots + (-1)^{\frac{q-1}{2}} \binom{q}{\frac{q-1}{2}} 2^{\frac{q-1}{2}} V_1.$$
q.e.d.

The results which will be proved now concern divisibility.

(14) If $m \geq 1$ and $k \geq 1$ then U_m divides U_{km}.

Proof: It is trivial if $k = 1$. If $k = 2$, by (6) $U_{2m} = U_m V_m$, so it is again true. Suppose that we have $k \geq 2$ and we already know that U_m divides U_{2m}, U_{3m}, up to $U_{(k-1)m}$. We proved in (8) that $U_{km} = U_{(k-1)m+m} = U_{(k-1)m} V_m - (-2)^m U_{(k-2)m}$. By assumption U_m divides $U_{(k-1)m}$ and also $U_{(k-1)m}$, hence U_m divides U_{km}.

From the beginning of the proof, U_m divides U_m and U_{2m}, so U_m divides U_{3m}, therefore U_m divides U_{4m}, etc. We conclude that U_m divides each U_{km} when $k \geq 1$.
q.e.d.

Eric. Please explain your last conclusion.

P.P. I do by reduction to an absurd. If there exists k such that U_m does not divide U_{km}, let us think that this k is the smallest possible. We know k is neither 1 nor 2. Also, U_m divides $U_{(k-1)m}$ and $U_{(k-2)m}$. We deduce that U_m divide U_{km}. This is a contradiction. My assumption that there exists k such that U_m does not divide U_{km} was not right.

Eric concluded:

Eric. Therefore U_m divides each U_{km}. Thanks for the convincing explanation. What is next? More divisibility properties?

P.P. Many more to come. They are the essential ingredients in the proof of the primality criterion for Mersenne numbers. First, I deduce some easy consequences of the relations just proved. Remember, if p is a prime then p divides each binomial coefficient $\binom{p}{k}$ where $k = 1, 2, \ldots, p-1$.

(15) Let $m \geq 1$, let p be an odd prime. Then $U_{mp} \equiv 12 U_m \pmod{p}$ and, taking $m = 1$, then $U_p \equiv 12^{\frac{p-1}{2}} \pmod{p}$.

Proof: We apply the relation (12) with q equal to the odd prime p. We wish to obtain a congruence modulo p. Now each coefficient $\binom{p}{1}, \binom{p}{2}, \ldots, \binom{p}{p-1}$ is congruent to 0 modulo p. What remains is the announced congruence $U_{mp} \equiv 12^{\frac{p-1}{2}} U_m \pmod{p}$. q.e.d.

(16) If p is an odd prime then $V_p \equiv 2 \pmod{p}$.

Proof: Now we apply the relation (13), replacing q by p. By Fermat's little theorem, $2^{p-1} \equiv 1 \pmod{p}$, so multiplying by 2, we have $2^p \equiv 2 \pmod{p}$. And we obtain the congruence $2 \equiv V_p \pmod{p}$. q.e.d.

One day, Lucas stood in front of the mirror and said (we imagine) to his image:

> *Lucas*: You seem to me more handsome than what my wife has ever said. And you're probably smarter than I. Something puzzles me about my beloved binary recurring sequences. If I pick any term, with some effort it is possible to know which are the prime factors of that term. Now listen to my question: Suppose I pick a prime p and I search for a term of the binary recurring sequence which is a multiple of the picked prime p. I wonder two things. Is it true that for every prime p there is a term which is a multiple of p? If for some reason it turns out that p divides a term, how small can the index n of the term be?

The mirror image answered:

> *Sacul*: Your question is most interesting. Let me try with the simplest sequences, first with the numbers L_n, which you have invented. We try $p = 3$ and then $p = 5$. We may replace the terms L_n by their least residues modulo p and we want to see if a least residue is equal to 0.
> Let $p = 3$; the least residues are
>
> $\qquad$ 1 1 0 1 1 2 0 2 2 1 0 ....
>
> You see that the least residues reappear as they started. Check the positions of the zeros. You have $L_2 \equiv 0$

(mod 3), $L_6 \equiv 0$ (mod 3), $L_{10} \equiv 0$ (mod 3) etc. So $p = 3$ divides $L_2, L_6, L_{10}, L_{14}, \ldots$.

Now let $p = 5$ to see if things happen in a similar way. The least residues modulo 5 are
$$2 \quad 1 \quad 3 \quad 4 \quad 2 \quad 1 \quad 3 \ldots.$$
This time, the least residues also reproduce themselves periodically and they are never equal to 0. This means that 5 never divides a number L_n.

Then I look at Fibonacci numbers, say $p = 5$, $p = 7$, or $p = 11$. I just write the least residues and it will be clear, as expected, that they reproduce periodically

$p = 5$: 0 1 1 2 3 0 3 3 1 4 0 4 4 3 2 0 2 2 4 1 0 1 1 2 ....

For $p = 5$ the period has 20 terms, 5 divides F_5, F_{10}, F_{15}, F_{20}, F_{25}, $F_{30} \ldots$

$p = 7$: 0 1 1 2 3 5 1 6 0 6 6 5 4 2 6 1 0 1 1 ....

The period now is 16 and 7 divides F_8, F_{16}, F_{24}, ...

$p = 11$: 0 1 1 2 3 5 8 2 10 1 0 1 1 ....

The period is 10 and 11 divides F_{10}, $F_{20}, \ldots$.

Sacul: After looking at these calculations, here is my advice:

(1) Mistrust the numbers L_n. The failure for $p = 5$ may indicate the possibility that many more primes are like 5, never dividing any L_n. You should try more primes like $p = 7, 11, 13, \ldots$. Beware.

(2) I see a possible theorem for Fibonacci numbers: Every prime p divides F_p, or F_{p-1} or F_{p+1}. You should prove it and decide for each prime p if p divides F_p, F_{p-1} or F_{p+1}.

Lucas: It was good to speak to you Sacul. You are so smart, I'll follow your advice.

Paulo listened to the mirror story, but had read "Alice Through the Looking Glass". He was more impressed with the strange fact that Fibonacci and Lucas numbers were not behaving in quite the same way.

Paulo. What happened after the conversation between Lucas and Sacul?

P.P. Lucas, who was not lacking in brains, studied binary recurring sequences like Fibonacci numbers beginning with 0 and 1, and proved the theorem suggested by Sacul.

Lucas' Theorem *for Fibonacci numbers.*

(1) *2 divides the Fibonacci numbers F_n with index n multiple of 3, and no others.*

(2) *5 divides the Fibonacci numbers F_n with index n multiple of 5 and no others.*

(3) *If p is a prime different from 2 and 5, then p divides F_n for all indices n multiples of $p - \left(\frac{5}{p}\right)$ where $\left(\frac{5}{p}\right)$ is the Legendre symbol, but it may happen that p divides also other Fibonacci numbers F_n.*

Eric. Ah! A very nice theorem. You only need to look at the index and you'll already find an index n such that p divides F_n. Good for lazy people.

Paulo. I agree that it is good. But you have also to agree it is not perfect. Lucas noted that "It may happen ... ta ta ta." A theorem should say what really happens.

P.P. At this point, there is a behavior of primes which is hard to explain. Just take primes p different from 2 and 5. Since there is one index n such that p divides F_n, then there is the smallest index n with this property. I give a name to this index. It is the smallest such that p appears as a factor of a Fibonacci number. So we call it the *index of appearance of p*. And we use some special notation for this number. Say $A(p)$, the A to remind us of the word "appearance". What happens is that $A(p)$ divides $p - \left(\frac{5}{p}\right)$, but may be smaller than $p - \left(\frac{5}{p}\right)$. Then p also divides all F_n where

n is a multiple of $A(p)$, but no other F_n. Just take $p = 13$. Then $A(13) = 7$, while $13 - \left(\frac{5}{13}\right) = 13 + 1 = 14$. This phenomenon has been carefully studied.

Eric. Will you tell us what is known?

P.P. That is *not* part of my debt, so I will not tell, as I wish to finish paying! Right now, I will explain what Lucas proved for the binary recurring sequences of U's:

(17) (a) U_n is even for all $n \neq 1$.
 (b) 3 divides U_n for all n multiple of 3 and for no others.
 (c) If $p \neq 2, 3$ is a prime, then p divides $U_{p - \left(\frac{3}{p}\right)}$.

Proof:

(a) This is trivial.

(b) We write the least residues of the U_n modulo 3

$$0 \quad 1 \quad 2 \quad 0 \quad 1 \quad 2 \quad 0 \ldots$$

These residues have period 3 and $U_n \equiv 0 \pmod{3}$ exactly when 3 divides n. In other words, 3 divides U_n if, and only if, 3 divides n.

(c) This is the important part of the theorem. The Legendre symbol $\left(\frac{3}{p}\right)$ (when $p \neq 3$) can be equal to 1 or to -1. We shall use the relation (11).

First assume that $\left(\frac{3}{p}\right) = -1$ and $n = p + 1$ in (11). We have the binomial coefficients

$$\binom{p+1}{1} = p + 1 \equiv 1 \pmod{p}$$

$$\binom{p+1}{3} = \frac{(p+1)p(p-1)}{1 \times 2 \times 3} \equiv 0 \pmod{p}$$

$$\binom{p+1}{5} = \frac{(p+1)p(p-1)(p-2)(p-3)}{1 \times 2 \times 3 \times 4 \times 5} \equiv 0 \pmod{p}$$

etc. Therefore

$$U_{p - \left(\frac{3}{p}\right)} = U_{p+1}$$

$$= \binom{p+1}{1} + \binom{p+1}{3} \times 3 + \binom{p+1}{5}$$

$$\times 3^2 + \cdots + \binom{p+1}{p} \times 3^{\frac{p-1}{2}}$$

$$\equiv 1 + 3^{\frac{p-1}{2}} \pmod{p}.$$

But, if you still remember, $3^{\frac{p-1}{2}} \equiv \left(\frac{3}{p}\right) = -1 \pmod{p}$ hence $U_{p-\left(\frac{3}{p}\right)} \equiv 0 \pmod{p}$, that is, p divides $U_{p-\left(\frac{3}{p}\right)}$.

Now we consider the case when $\left(\frac{3}{p}\right) = 1$. We shall use the relation (11) with $n = p - 1$. We have

$$\binom{p-1}{1} = p - 1 \equiv -1 \pmod{p}$$

$$\binom{p-1}{3} = \frac{(p-1)(p-2)(p-3)}{1 \times 2 \times 3} = (-1)^3 \frac{1 \times 2 \times 3}{1 \times 2 \times 3} \equiv -1 \pmod{p}$$

$$\binom{p-1}{5} = \frac{(p-1)(p-2)(p-3)(p-4)(p-5)}{1 \times 2 \times 3 \times 4 \times 5}$$

$$\equiv (-1)^5 = -1 \pmod{p},$$

etc. Hence

$$U_{p-\left(\frac{3}{p}\right)} = U_{p-1}$$

$$= \binom{p-1}{1} + \binom{p-1}{3}$$

$$\times 3 + \binom{p-1}{5} \times 3^2 + \cdots + \binom{p-1}{p-2} \times 3^{\frac{p-3}{2}}$$

$$\equiv -1 - 3 - 3^2 - \cdots - 3^{\frac{p-3}{2}} \pmod{p}.$$

Do you remember this one:

$$1 + a + a^2 + \cdots + a^k = \frac{a^{k+1} - 1}{a - 1}?$$

You do? OK! I apply it with $a = 3$, $k = \frac{p-3}{2}$ and I conclude that the preceding expression is equal to $-\frac{3^{\frac{p-1}{2}} - 1}{3 - 1} \equiv 0 \pmod{p}$ because

$$3^{\frac{p-1}{2}} \equiv \left(\frac{3}{p}\right) \equiv 1 \pmod{p}.$$

Thus p divides $U_{p-\left(\frac{3}{p}\right)}$ also in this case and the proof is concluded. q.e.d.

(18) Let p be a prime different from 2 and 3. Let $m \geq 1$, $e \geq 1$, $f \geq 1$ and assume that p^e divides U_m. Then p^{e+f} divides $U_{p^f m}$.

Proof: We prove the result when $f = 1$, that is, we show that p^{e+1} divides U_{pm}. For this purpose we write (12), with q replaced by p:

$$12^{\frac{p-1}{2}} U_m^p = U_{pm} - \binom{p}{1}(-2)^m U_{m(p-2)} + \binom{p}{2}(-2)^{2m} U_{m(p-4)}$$

$$- \cdots + (-1)^{\frac{p-1}{2}} \binom{p}{\frac{p-1}{2}}(-2)^{\frac{p-1}{2} \times m} U_m.$$

By assumption, p^e divides U_m, hence, by (14) also p^e divides $U_{n(p-2)}, U_{m(p-4)}, \ldots$, therefore p^{e+1} divides $12^{\frac{p-1}{2}} U_m^p$ and also all the terms, following the first one, of the right-hand side of the preceding expression. It follows that p^{e+1} must divide U_{pm}.

The same reasoning may be repeated, taking pm in place of m and $e+1$ in place of e. We deduce that if p^e divides U_m, then p^{e+2} divides $U_{p^2 m}$. By repetition, the proof may be completed. q.e.d.

Paulo. What is next?

P.P. I will consider the following situation. If $N > 1$, if there exists $n > 1$ such that N divides U_n, then there exists the smallest integer, which shall be denoted by $A(N)$, such that N divides $U_{A(N)}$.

Remember, for Fibonacci numbers, $A(p)$ denoted the index of appearance of the prime p. For the sequence of U's, by (17) every prime p divides some term U_n, namely 2 divides U_2, so $A(2) = 2$. 3 divides U_3, so $A(3) = 3$ and if $p \neq 2, 3$, then p divides $U_{p-\left(\frac{3}{p}\right)}$, so $A(p) \leq p - \left(\frac{3}{p}\right)$.

(19) Let $n > 1$, let N be odd. Then N divides U_n if and only if $A(N)$ divides n.

Proof: Assume first that $A(N)$ divides n. Then $U_{A(N)}$ divides U_n, by (14). But N divides $U_{A(N)}$, so N divides U_n.

Now we assume that N divides U_n, so $A(N) \le n$. Let $n = q \times A(N) + r$, where $0 \le r < A(N)$. We wish to show that $r = 0$, so $A(N)$ divides n. We use (9):

$$U_n V_{q \times A(N)} - U_{q \times A(N)} V_n = (-2)^{q \times A(N)} U_r.$$

By assumption N divides U_n, also by (14), N divides $U_{q \times A(N)}$, hence N divides $(-2)^{q \times A(N)} U_r$. But N is odd, hence N divides U_r. From $r < A(N)$ it follows that $r = 0$, as it was required to prove.

q.e.d.

The Proof of the Primality Criterion for Mersenne Numbers

Eric. Papa Paulo, you have been numbering the facts about binary recurring sequences. You have defined these sequences and deviated from your main purpose to talk about rabbits, Fibonacci numbers and Lucas numbers. Then, you considered the sequences of U's and V's which had to be studied together. First, you proved several relations which these numbers satisfy. Then you looked at properties involving divisibility. It was a nice build up. I was quite amazed to see how many steps, proven systematically in the right order, turned out to be required. No doubt, the most important fact — in my view — was that if p is a prime, different from 2 and 3, then p divides $U_{p - \left(\frac{3}{p}\right)}$.

P.P. You have analyzed the development quite well. I find the arithmetical theory of binary recurring sequences a highly interesting area of number theory. I did not mention it, but there remain many challenging problems, some of which have been solved, but many more unsolved.

Paulo. One day, will you mention mysteries involving prime numbers and Fibonacci numbers?

P.P. Indeed I will, but now I'm heading towards the proof of the criterion.

Paulo. "Heading towards" means that there is more preparation?

P.P. Yes, some very specific preparation. First I will define the number $T(N)$ where $N = \prod_{i=1}^{s} p_i^{e_i}$, I assume that 2 and 3 do not divide N, $s \geq 1$, each $e_i \geq 1$. The number $T(N)$ is defined to be

$$T(N) = \frac{1}{2^{s-1}} \prod_{i=1}^{s} p_i^{e_i-1}\left(p_i - \left(\frac{3}{p_i}\right)\right).$$

If $N = p$ is a prime $p \neq 2, 3$, then $s = 1$, $p_1 = p$, $e_1 = 1$, so $T(p) = p - \left(\frac{3}{p}\right)$.

More generally:

(20) If N is not a multiple of 2 or 3:
  (a) If $N = p^e$, with $e \geq 2$ then $T(N) < N - 1$ when $\left(\frac{3}{p}\right) = 1$ and
$$T(N) > N + 1 \text{ when } \left(\frac{3}{p}\right) = -1.$$
  (b) If $s \geq 2$, then $T(N) < N - 1$.

Proof:

(a) $T(N) = p^{e-1}\left(p - \left(\frac{3}{p}\right)\right) = N - p^{e-1}\left(\frac{3}{p}\right).$

If $\left(\frac{3}{p}\right) = 1$, then $T(N) = N - p^{e-1} \leq N - p < N - 1$.

If $\left(\frac{3}{p}\right) = -1$, then $T(N) = N + p^{e-1} \geq N + p > N + 1$.

(b) Let $s \geq 2$, so

$$T(N) = \frac{1}{2^{s-1}} \prod_{i=1}^{s} p_i^{e_i-1}\left(p_i - \left(\frac{3}{p_i}\right)\right) \leq \frac{1}{2^{s-1}} \prod_{i=1}^{s} p_i^{e_i-1}(p_i + 1)$$

$$= \frac{N}{2^{s-1}} \prod_{i=1}^{s}\left(1 + \frac{1}{p_i}\right) = 2N \prod_{i=1}^{s} \frac{1}{2}\left(1 + \frac{1}{p_i}\right) < N - 1,$$

because $N > 5$. q.e.d.

(21) If N is not a multiple of 2 or 3, then N divides $U_{T(N)}$.

Proof: With the notations being used, each prime $p_i \neq 2, 3$. By (18) $p_i^{e_i}$ divides $U p_i^{e_i-1}(p_i - \left(\frac{3}{p_i}\right))$.

Let ℓ be the least common multiple of the numbers $p_i^{e_i-1}(p_i - \left(\frac{3}{p_i}\right))$, for $i = 1, \ldots, s$.

By (14), $p_i^{e_i}$ divides U_ℓ. Hence $N = \prod_{i=1}^{s} p_i^{e_i}$ divides U_ℓ. But ℓ is equal to 2 times the least common multiple of the numbers $\frac{1}{2}p_i^{e_i-1}(p_i - (\frac{3}{p_i}))$. So ℓ divides

$$2 \prod_{i=1}^{s} \frac{1}{2} p_i^{e_i-1}\left(p_i - \left(\frac{3}{p_i}\right)\right) = \frac{1}{2^{s-1}} \prod_{i=1}^{s} \frac{1}{2} p_i^{e_i-1}\left(p_i - \left(\frac{3}{p_i}\right)\right) = T(N).$$

By (14) U_ℓ divides $U_{T(N)}$, so N divides $U_{T(N)}$. q.e.d.

P.P. We are coming close to the primality test.

(22) Let n be odd, $n \geq 3$, let $N = M_n = 2^n - 1$. If N divides U_{N+1}, but N does not divide $U_{\frac{N+1}{2}}$, then N is prime.

Proof: First note that N is odd and since n is odd, $N = 2^n - 1 \equiv 1 \pmod{3}$, so 3 does not divide N. By (21) N divides $U_{T(N)}$, so there exists the smallest integer $A(N)$ such that N divides $U_{A(N)}$. By (19) $A(N)$ divides $T(N)$ and also $A(N)$ divides $N+1 = 2^n$, but by (19), $A(N)$ does not divide $\frac{N+1}{2} = 2^{p-1}$. Hence $A(N) = 2^n = N+1$.

If N is not prime and N has at least two distinct prime factors, then by (20) $N+1 \leq T(N) < N-1$, which is absurd. If $N = p^e$ with $e \geq 2$ and p prime, by (20) $N+1 < T(N)$ so $(\frac{3}{p}) = -1$ and $T(N) = p^{e-1}(p - (\frac{3}{p})) = p^{e-1}(p+1) = p^e + p^{e-1} = (N+1) + (p^{e-1} - 1)$. But $N+1$ divides $T(N)$, so $N+1 = p^e + 1$ divides $p^{e-1} - 1$. This is a contradiction. Therefore N is a prime. q.e.d.

P.P. Thanks for your patience. Now I will again state the test of primality for Mersenne numbers and give its proof!

(23) **Primality test for Mersenne numbers.** Let n be odd, $n \geq 3$, let $N = M_n = 2^n - 1$. Then N is prime if and only if N divides $V_{\frac{N+1}{2}}$.

Proof: We assume first that N is a prime. By (7)
$$V_{\frac{N+1}{2}}^2 = V_{N+1} + 2(-2)^{\frac{N+1}{2}} = V_{N+1} - 4(-2)^{\frac{N-1}{2}} = V_{N+1} - 4(\frac{-2}{N}).$$
But $N \equiv 7 \pmod 8$, hence $(\frac{-2}{N}) = (\frac{-1}{N})(\frac{2}{N}) = -1$. Therefore $V_{\frac{N+1}{2}}^2 = V_{N+1} + 4$.

By (10) $2V_{N+1} = V_N V_1 + 12 U_N U_1 = 2V_N + 12 U_N$, so $V_{N+1} = V_N + 6U_N \equiv 2 + 6 \times 12^{\frac{N-1}{2}} \equiv 2 + 6 \times \left(\frac{12}{N}\right) \equiv 2 + 6\left(\frac{3}{N}\right) = 2 - 6 \equiv -4 \pmod{N}$ by virtue of (16) and (15). Therefore $V_{N+1} + 4 \equiv 0 \pmod{N}$. So N divides $V_{\frac{N+1}{2}}^2$, hence N divides $V_{\frac{N+1}{2}}$.

Now we prove the converse. Assume that N divides $V_{\frac{N+1}{2}}$. By (6) N divides U_{N+1}. By (5) $V_{\frac{N+1}{2}}^2 - 12 U_{\frac{N+1}{2}}^2 = 4(-2)^{\frac{N+1}{2}}$. If N divides $U_{\frac{N+1}{2}}$ then N divides $4 \times (-2)^{\frac{N+1}{2}}$. This is impossible because N is odd. So N does not divide $U_{\frac{N+1}{2}}$. By (22), N is a prime and the proof is completed, completed, completed. q.e.d.

Eric. This proof is complete, but the test for Mersenne primes which you stated earlier was in terms of numbers S's. This time, you didn't even mention these numbers.

P.P. Sorry for my omission. The numbers S's are closely related to the numbers V's.

I recall the definition of the numbers S's:

$S_0 = 4$, $S_1 = S_0^2 - 2$ and for every $k \geq 2$, $S_k = S_{k-1}^2 - 2$.

(24) For every $k \geq 0$:

$$S_k = \frac{V_{2^{k+1}}}{2^{2^k}}.$$

Proof: The proof is by the method of induction. For $k = 0$, $\frac{V_2}{2} = \frac{8}{2} = 4 = S_0$. Now we assume that for all integers $0, 1, \ldots, k-1$ the statement has already been proved. We shall prove that it is also true for k, hence it is never false. By definition

$$S_k = S_{k-1}^2 - 2 = \left(\frac{V_{2^k}}{2^{2^{k-1}}}\right)^2 - 2 = \frac{V_{2^k}^2}{2^{2^k}} - 2$$

$$= \frac{V_{2^{k+1}} + 2^{2^{k+1}}}{2^{2^k}} - 2 = \frac{V_{2^{k+1}}}{2^{2^k}},$$

as it was required to prove. q.e.d.

Eric. The relation between the V's and the S's is very simple. Why did you not stick to the V's? By the way, why did you not stick to the U's and use (22)?

P.P. Simple answer, it is easier to calculate the S's than the U's or the V's. In terms of the numbers S's, the primality criterion becomes:

(25)   The Mersenne number M_n is a prime if and only if M_n divides S_{n-2}.

Proof: From (23), M_n is prime if and only if M_n divides $V_{2^{n-1}} = 2^{2^{n-2}} \times S_{n-2}$, or equivalently M_n divides S_{n-2}.   q.e.d.

Paulo and Eric were happy that my debt was paid. If they were bankers, they would immediately try to convince me to be hooked again. I will not.

Paulo. I am impressed with your determination to pay your debt. You mentioned so many exciting facts. I am especially attracted by the Fibonacci and the Lucas numbers. Is there any easy book to read when vacation arrives? Something relaxing and not too demanding on my brain.

P.P. I can indicate one book which is exactly what you desire. The author is V. E. HOGGATT, Jr., the title is *"Fibonacci and Lucas Numbers"*. It is thin, only 92 pages. You may take the book to the beach, but don't forget to wear good sunglasses, because of the ultraviolet sunrays.

P.P. And I will envy both of you, but you are smart and nice guys. But, for the moment, it is not yet vacation time. From your enthusiasm, I understand that you would be happy to read about Fibonacci. Here are some notes on him, and some on Lucas, too.

Notes About Leonardo Pisano, known as Fibonacci (circa 1170–1250)

Leonardo Pisano, the son of Guillelmo Bonacci, was born in Pisa (Italy) and became known as "Fibonacci". With his family he spent his youth in Bejaia (Algeria), traveling extensively in the Mediterranean countries for commercial purposes. Being very gifted in mathematics, he learned the use of Indo-Arabic numerals and appreciated its advantages over the Roman system in use in Europe. Upon returning to Pisa, he published in 1202 the book *"Liber Abaci"*, in which he explained the use of the new numerals.

The book contained numerous solved arithmetic problems, including the problem of rabbits, which led to the sequence of numbers now known as the *Fibonacci numbers*.

The book *"Practica Geometriae"* concerns problems in elementary geometry and contains, among other results, the expression for the length of the side of a regular inscribed pentagon, or decagon, in terms of the radius of the circle. The most significant book written by Fibonacci was the *"Liber Quadratorum"*, published in 1225. It deals with squares and problems leading to quadratic equations. Fibonacci gives a method to construct Pythagorean triples and he shows that selected quadratic equations in two unknowns have no solutions in positive integers.

Fibonacci enjoyed a well-deserved reputation during his life; his work represented a considerable advance over what had been known in the area of arithmetic. It is generally agreed that the number theory results of Fibonacci were the most important since the time of Diophantus and before Fermat.

Notes About François Edouard Anatole Lucas (1842–1891)

Edouard Lucas was born in Amiens (France), where he studied to become a mathematics teacher. He occupied successive positions in the prestigious Parisian lycées Saint-Louis and Charlemagne. Lucas' work was in number theory and concomitantly he produced a substantial number of papers in mathematical recreations, gathered in four volumes. In number theory, Lucas investigated the sequence of Bernoulli numbers and applied the symbolic method to derive numerous relations between Bernoulli numbers. His most important contributions were towards primality testing and factorization, using binary recurring sequences, now called *Lucas sequences*. He wrote a long seminal paper on the subject, which was published in 1878 in the *American Journal of Mathematics*. Lucas' primality criterion for Mersenne numbers allowed him to prove that $2^{127} - 1$ is a prime number, at that time the largest known prime.

Only volume I of the treatise *"Théorie des Nombres"* (planned for four volumes) was published. A freak accident cut Lucas' life short at age 49. At a banquet, a piece of a broken plate injured his cheek and caused an infection, from which he died. Food caused mathematics losses.

Notes About Pythagoras of Samos (circa 569 BCE–circa 475 BCE)

Pythagoras was born in Samos, an island in the Ionian Sea. At a young age, Pythagoras showed an intense interest in mathematics. He met Thales of Miletus and followed the lectures on geometry of Thales' disciple Anaximander. Pythagoras spent a long period in Egypt, where he became interested in and influenced by their religious and philosophical ideas. During the war against the Persians he was taken as a prisoner to Babylon, where he stayed for several years. After his liberation, he left Samos for Cotrone, a town in the southeast of Italy. There he founded his school, which was in fact a community living under strict rules, with lofty and noble purpose. Music, poetry, philosophy, astronomy, but mostly mathematics were cultivated.

Pythagoras is usually considered the first pure mathematician. Due to their rules of secrecy, the Pythagoreans left no writings, but many of their discoveries can be ascertained by oral tradition. The famous "Pythagorean theorem" taught in schools was in fact known by the Babylonians. It is agreed, however, that the first proof is due to Pythagoras. More theorems of geometry were discovered and proved by the Pythagorians, like the construction of tetrahedron, cube and octahedron, the values $(2n-4) \times 90°$ of the sum of angles of a convex polygon of n sides. Pythagoras, who played the lyre quite well, realized that strings produce harmonic sounds when the ratio of their lengths is an integer. For Pythagoras' school, reality was mathematical in its nature and ratios of integers describe all quantitative measurements. The discovery that the hypotenuse of a right-angled triangle with sides measuring 1 was not a rational number, represented a shattering blow to the basic beliefs of the Pythagoreans.

Notes About Blaise Pascal (1623–1662)

Born in Clermont-Ferrand (France), Pascal showed an acute mathematical intelligence very early. At age 14, he already participated in meetings of the Académie Parisienne, founded by Mersenne. Pascal followed the ideas of Desargues and used methods of projective geometry in an important treatise on conics. His book on the arithmetical triangle contains a variety of results on combinatorial analysis, which even though not all original, were obtained in a systematic way. Together with Fermat, Pascal is the recognized founder of the theory of probability. Pascal also devoted his attention to the invention of a machine to add and subtract numbers. In physics, he had shown interest for the study of vacuum and barometric pressure. Beyond mathematics and science Pascal is the author of many publications, including the classical "*Pensées*". He also entered into numerous controversies concerning philosophical and religious arguments.

18

Money and Primes

Eric. I like both money and primes. I am surprised that they may team up together!

Money

P.P. They do, and what I'll say is not just to please you. Listen to what the Prince wrote once in 1801 — when he was still a teenager.

"*The problem of distinguishing prime numbers from composite numbers and of resolving the latter into their prime factors is known to be one of the most important and useful in arithmetic. The dignity of the science itself seems to require that every possible means be explored for the solution of a problem so elegant and so celebrated.*"

Paulo. Who was this teenage Prince? I would have said the same thing, but only my mother calls me Prince.

P.P. He was Prince Gauss and his realm was going to be mathematics. Gauss made the statement in article 329 of his book opera magna "*Disquisitiones Arithmeticae*". It is a fact that the Greeks and many predecessors of Gauss, as I have mentioned, were already interested in recognizing if a given number N is a prime. The task for the researchers was the following:

To devise a method, test, procedure, algorithm, or program — whatever name you want to use, consisting of a list of successive arithmetical operations to perform on the given number N and on numbers obtained in preceding steps, until reaching a conclusion, which you can call an output, which is either "N is prime" or "N is composite".

Eric remembered our previous discussions and said:

Eric. We already know an algorithm for this purpose. Just try Euclidean division by each integer k which is less than $\sqrt{N}$ — if none divides N then N is a prime. In fact, it suffices to try division by the primes $p < \sqrt{N}$, if one happens to know them all.

P.P. This is absolutely obvious. We may say that the problem of recognition if a given number N is prime is decidable in a finite number of steps. I now want to go somewhat deeper into the matter. What do I mean by "steps"?

Papa Paulo continued:

P.P. I should have said arithmetical operations instead of steps. Now it is clear that if I have to add $5 + 7$ it is less work than to add $538 + 7718$ and even less work than when I add a number with 330 digits and one with 520 digits. In all the above cases, it is just one arithmetical operation, but the more digits, the more operations with the digits that have to be done.

Eric. Once you called "bit operations" those done with digits. Anyway, what you said is obvious.

P.P. People are interested in knowing in advance what to expect from an algorithm to test the primality of a number N. How many bit operations will be needed? How long it will take to perform the operations? How much it will cost? The number of bit operations has to depend on the number of digits of N. I write $d(N)$ for the number of digits of N. It is sometimes possible — not easy but possible — to determine that the number of required bit operations to treat N cannot be more than $(d(N))^e$, where e is a positive integer, the same for all N.

Eric. Make yourself clear.

P.P. OK. You must be an expert in analyzing the complexity of an algorithm. This means that you keep track of how many bit operations you need to perform in the successive steps of the algorithm when you start with a number N with $d(N)$ digits. For example, the algorithm is such that the number of bit operations to test an arbitrary number N is at most $(d(N))^5$.

Eric. I understand what is going on, but only superficially, because I am not and I do not intend to become an expert in complexity of algorithms.

P.P. If an exponent e (like $e = 5$) has been found for the algorithm A, it is said that A is of *polynomial type*. This is the best — in fact the only — desirable situation.

Eric. What else could happen?

P.P. In many known algorithms bad things like this may happen: the number of bit operations (apart from exceptional values of N — the friendly N's) is not less than $\sqrt{N}$.

Eric. What is so bad about $\sqrt{N}$?

P.P. Let me concoct an example to illustrate what may happen. Say, you have a computer which is able to perform 6×10^6 bit operations per second.

Just a number to support the calculations. Say, you have an algorithm A for which the number of bit operations to test an arbitrary number N is at most $(d(N))^5$. And you also have an algorithm B which needs at least $\sqrt{N}$ bit operations to treat unfriendly numbers N. Now pick an unfriendly number N with 300 digits. (I stress that most numbers are unfriendly for any algorithm.)

Algorithm A.
(Number of bit operations needed to test N)
$\geq 300^5 = (3 \times 10^2)^5 = 3^5 \times 10^{10}$.
(Number of seconds needed to perform the bit operations)
$\leq \frac{3^5 \times 10^{10}}{6 \times 10^6} = \frac{3^5}{6} \times 10^4$.
(Number of minutes needed to perform the bit operations)
$\leq \frac{3^5}{6^2} \times 10^3$.

(Number of hours needed to perform the bit operations)
$\leq \frac{3^5}{6^3} \times 10^2$.
(Number of days needed to perform the bit operations)
$\leq \frac{3^5}{4 \times 6^4} \times 10^2 = \frac{3}{2^6} \times 10^2 = \frac{75}{16} < 5$.

Algorithm B.
(Number of bit operations needed to test N)
$> \sqrt{N} > 10^{149}$ because $10^{299} \leq N < 10^{300}$.
(Number of seconds needed to perform the bit operations)
$> \frac{10^{149}}{6 \times 10^6} = \frac{1}{6} \times 10^{143}$.
(Number of minutes needed to perform the bit operations)
$> \frac{1}{6^2} \times 10^{142}$.
(Number of hours needed to perform the bit operations)
$> \frac{1}{6^3} \times 10^{141}$.
(Number of days needed to perform the bit operations)
$> \frac{1}{4 \times 6^4} \times 10^{141}$.
(Number of years needed to perform the bit operations)
$> \frac{1}{4^2 \times 6^4} \times 10^{139}$.
(Number of centuries needed to perform the bit operations)
$> \frac{1}{4^2 \times 6^4} \times 10^{137} \, 7 > 10^{132}$.

P.P. Just to give you an idea of the indescribable length of the algorithm, think that the astrophysicists evaluate that the Big Bang occurred some 14×10^9 years ago.

Paulo and Eric (*in harmony*). Absolutely unthinkable. We would never suspect that a bad algorithm could require such a long time to be performed.

P.P. For this reason it is essential to invent algorithms of polynomial type, or at least very close to be of polynomial type. The cost C to perform an algorithm is directly proportional to the required number of bit operations, and inversely proportional to the speed of the computer. Progress will come from improved hardware, that is, faster computers with more memory. But progress is also dependent to a large extent on the invention of clever algorithms. The search for these algorithms mobilizes deeper and deeper facts from number theory, more specifically from algebraic number theory and arithmetic geometry. Progress in primality testing needs and fosters progress in certain areas of number theory. And do not forget, large numbers are hard to manipulate, even with computers. Experts in computational number theory invent unbelievable tricks to deal with big numbers, like the fast Fourier transform.

Eric. I am overwhelmed by the extent the subject of prime numbers has developed.

P.P. Governments of powerful countries, through their military forces or banking institutions, are supporting the research on prime numbers. Thousands of brains serve this effort, to learn more and more about prime numbers. All this abundant funding aims at immediate direct applications.

Immediate direct applications of number theory! Who would dream of it, some 40 years ago? Poor number theory, the Queen relegated (or raised?) to be the object of a courtship inspired by greed and necessity, not by awe.

Am I a Good Guy?

P.P. Listen to this fictional story.

Like everyone, I seek approval. I would like to know if I am a good guy. Most of my acquaintances look at me with indifference. For many, I am unfriendly. Only a very few consider that I am friendly. I mistrust judgments which are not supported by unquestionable facts. Opinions which are vague or imprecise do not suffice. I would like to know, without undue delay, if I am a good guy.

It is about the same with numbers. Mister N wants to know if he is a "prime guy". He would like to be tested. It may be a friendly or an impartial test, which would better be justified and with a fast verdict: N wants to know if he is a "prime guy".

Paulo. What a fancy story. Numbers don't ask to be tested. From your story, I understand that there are many kinds of tests.

Eric. Yes, Paulo. We already saw the test by trial division and the special tests for Mersenne numbers.

Paulo. Papa Paulo, are you going to talk about other tests?

P.P. This is what I want to discuss. Primality tests can be considered from various points of view.

$$\begin{cases} \text{Tests for numbers of special form} \\ \text{Tests for generic numbers} \end{cases}$$

$$\begin{cases} \text{Tests with full justification} \\ \text{Tests with justification based on conjectures} \end{cases}$$

$$\begin{cases} \text{Deterministic tests} \\ \text{Probabilistic tests} \end{cases}$$

Paulo. I like this classification. Will you tell us about all these kinds of tests?

P.P. I would have liked to do it, but I cannot.

Eric. Papa Paulo, you know it all. Why not?

P.P. Apart from what I have already explained, most tests require a far greater knowledge of number theory than what I could manage to discuss with you. To acquire the necessary background and see these tests in detail, you would need to attend university courses specially dedicated to the topic, coupled with courses in computational number theory. But I can still make comments to satisfy your curiosity.

Eric. Let us hear you, Papa Paulo.

P.P. Tests for numbers of special form: If it happens that $N - 1$ (or $N + 1$) has known prime factors, it was possible to devise primality tests for N which run in polynomial time. We have seen the tests for Fermat numbers $F_n = 2^{2^n} + 1$ and Mersenne numbers $M_q = 2^q - 1$. People must have invented tests for similar kinds of numbers. But, for given primes $p_1, \ldots, p_k$ there is only a negligible proportion of numbers N such that the prime factors of $N - 1$ (or $N + 1$) are among $p_1, \ldots, p_k$. These tests are valuable but of a limited scope.

Paulo. What is a "generic" number?

P.P. Nothing special, it is just any number. It may or may not have special properties, but no use will be made of these properties.

Paulo. Tests for generic numbers — anything particular about these tests?

P.P. They may be applied to any number irrespective of the knowledge of prime factors of $N-1$, or $N+1$, or anything special for the number. Of course, if one recognizes that N has friendly properties, as I mentioned earlier, then the best is to apply specially adapted tests. But for a random number ...

Eric. What is a "random number"?

P.P. To get a random number of 100 digits say, you turn a roulette with digits 0 to 9 successively 100 times.

Each time, it points to a digit $0, 1, 2, \ldots, 9$; you write the number N so obtained. There is a certain proportion of numbers — namely those which are even multiples of $2, 3, \ldots,$ and those with small prime factors — which are easily identified and discarded. The numbers N which are not disqualified, have no particular property, so they are tested by the methods designed for generic numbers.

Eric. Can you tell which are the most used, or at least say why, and tell us the names of the most used methods?

P.P. This is hard. Why? The primality tests used to deal with generic numbers need sophisticated mathematics both in their justification and for calculation. One such test is the APR test (not invented in April by Paulo Ribenboim); APR are the initials of Adleman, Pomerance and Rimely, who invented the test. The mathematics involves algebraic numbers — which I will not try to explain. The running time is not far superior to polynomial time.

Another test for generic numbers is based on elliptic curves (no explanation given!). To test the number N one has to find — somewhat by luck — an elliptic curve which will be appropriate to test the number. After the random choice of the curve, the test runs in polynomial time, so people say the elliptic curve test runs in "random polynomial time".

Eric. Papa Paulo, I know that we do not have the background to understand how these tests are justified and implemented. But can you tell us things like the size of numbers which can be tested?

P.P. This keeps increasing. I have not looked at the most recent performances. The APR method has tested in about 10 minutes the primality of numbers with 200 digits. The method using elliptic curves has been outperforming the APR test. Numbers with more than 500 digits have been found to be prime by the elliptic curves method.

Eric. If you come to me with a number with over 5000 digits (which is already difficult to contemplate) and you say: "Eric, this number has been shown to be a prime by the elliptic curve method," why should I believe you? The possibility of computational error exists. What about the guarantee of the assertion?

P.P. We treat the number with the method, three times, with different computers. Each time the output was "prime".

Eric. Is this enough to be confident?

P.P. Actually the calculations come with a "certification" of primality; it is a sequence of numbers which allows double checking. Eric, your point is important. To guarantee that computations involving zillions of bit operations are correct we repeat them with other computations. If the results coincide we are confident that the output is correct.

Paulo. And what about the justification of tests?

P.P. The tests with full justification are exactly what is said. Mathematicians propose a test which is based on known theorems and sometimes also on theorems expressly proved to justify the test.

Paulo. I am a bit puzzled. I thought that all one does in mathematics has to be supported by proofs, and you make a distinction between "tests fully justified" and "tests with justification based on conjectures". If I remember correctly, a conjecture is a statement which one believes to be true, but no one has succeeded in finding a proof.

P.P. This is exactly so. I will explain how it works. Mathematicians create a test T. At a certain point, the justification of T is based on the assumption that the conjecture C is true. We apply test T to the number N. And we proclaim its output "prime" or "composite" as if it were right. If it is, great. If it is not right, great again, because that would show that the conjecture is false, despite what everyone believed.

Eric. Dangerous way of doing mathematics.

P.P. Then prove me wrong!

After this blatant illustration of scientific intimidation, Eric asked:

Eric. Which is the name of the supporting conjecture?

P.P. It is a strong generalized Riemann Hypothesis.

This was the end of the discussion. The intimidation was total. Paulo asked:

Paulo. Tell us about deterministic and probabilistic tests.

P.P. I prefer to begin with a story. Suppose you are in a big city. The GPS has not yet been invented and you want to visit the Science Museum. You ask directions from a person you meet in the street. He is friendly and patient.

> *Person*: It is a little bit difficult to explain. I can do it if you have a piece of paper and a pen.

I have both, so the person draws a map with streets and gives very clear directions:

> *Person*:
>
> (1) Follow Avenue A up to the third street.
> (2) Turn to the right on a one-way street, and follow it to the end.
> (3) You will arrive at a big plaza, and across there will be the Science Museum, the second large building from the right.

I said "Thank you very much for your help," and thought to myself, there is no way I can go wrong.

Paulo. Why are you talking about the Science Museum? I wanted to listen to what you have to say about deterministic tests.

P.P. A deterministic test is like in my story. You already know, without any doubt, what to do. Now I tell another story which is like a probabilistic test. As before, I ask directions to go to the Science Museum. This time the man in the street is uncertain. This is what he said:

(1) Take the wide avenue at the right. There are about 12 streets along the way. You have to take one, but I don't remember which one. Some streets lead nowhere, but only one leads to a large plaza. Try one after the other until you find the good one which ends in the plaza.

(2) There you'll see about a dozen buildings, one of which is the Science Museum, but I cannot tell you which one — just try!

This time I thought to myself: I may be lucky at once, or only after trying many side streets. The probability of being right is 1 in 12. When I arrive at the plaza, after a few more trials, I will identify the Science Museum. I hope that after all this trouble the Museum will be interesting and will display an (almost) perpetual mobile.

Papa Paulo continued:

P.P. In a probabilistic test — as in any test — you follow the instructions. But you are allowed choices at various steps, with only a measured probability you may make the right choice which leads to an output. The output may be that the number N shares with the primes certain properties. And only a small percentage of numbers may be composite and have those properties of primes. So the number N is declared to be a "probable prime". The qualification "probable" reflects a degree of ignorance. After all, a number is either "prime" or "composite".

Paulo. Your ways are strange.

P.P. Remember, once I told you about the number $341 = 11 \times 31$. It is composite, but shares with the primes a certain property. This same idea has been much explored and quicky gives the outputs "probable prime" or "composite". For probable primes more work is needed to establish the primality of the number. But without further ado, we may safely believe that probable primes so obtained are, in fact, primes.

Eric. I see, faith comes into primality testing.

A silent pause, then Paulo inquired:

Paulo. About the running time for tests, you said that the special tests for Fermat numbers and for Mersenne numbers run in polynomial time. But for generic numbers the APR test runs in "almost polynomial time" (whatever that means) and the test with elliptic curves runs in random polynomial time. I wonder: Is there any test which runs in polynomial time for generic integers? It should also be deterministic (who wants probable primes?) and fully justified (I prefer not to appeal to conjectures). Anything like this?

P.P. Paulo, congratulations. You did not solve the problem which had been central in the area. (But you asked!) The problem has been solved.

Paulo. Why did you not tell us earlier? We could have avoided all kinds of tests.

P.P. Agrawal, Kayal and Saxena proved in 2002 that there is a deterministic fully justified primality test for generic integers which runs in polynomial time. This has solved the outstanding problem. The running time is a relatively high power of the number of digits of N. Until some improvement is made it is more practical to use the APR test or the elliptic curves test.

Titanic Primes and Megaprimes

P.P. In the early eighties, a prime with 1000 or more digits was hard to pinpoint. Even though we know that there are primes of arbitrarily large size, it was a sort of a noteworthy event when someone found a specific prime number with more than 1000 digits. A prime with 1000 or more digits was called a *titanic* prime. At first, very few were known. With progress in primality testing and computational methods, more and more titanic primes were identified. A friend told me about lists of titanic primes. In 1983 the list had 110 numbers, in 1988, there were 581 titanic primes of which 170 with more than 2000 digits. At the end of 2002 a list with the 5000 largest known primes was made available in an internet site; each one of these primes had at least 30 000 digits. This witnesses how fast has been the progress in prime recognition.

Paulo. Any special name for primes with 1 000 000 or more digits?

P.P. We call them *megaprimes*. The six largest known Mersenne primes are megaprimes, in fact, the only ones known today. The Mersenne prime
$$M = 2^{57885161} - 1$$
is the largest prime known today. It has 17425170 digits.

The story with megaprimes will be like for titanic primes. Today, who cares about titanic primes? More and more megaprimes will be discovered.

Eric. I have to confess my admiration for such technical prowess. How do people choose the numbers to be tested?

P.P. They have to be friendly, usually of special form like $k \times 2^n \pm 1$, $k^2 \times 2^n \pm 1$, $k \times 10^n \pm 1$, etc. For these numbers there are special primality tests which run in polynomial time. Yet, as seen for Mersenne numbers, for large values of n, teams of people work together.

Curious Primes

We are curious, we like to play with numbers and when we find primes whose digits show nice patterns, we find happiness.

P.P. The number 35853 is an example of a *palindromic number*. It is the same number, whether you read it from left to right or right to left.

Paulo. Big deal. Anybody can write palindromic numbers.

P.P. True, but the game is to find palindromic primes. With one digit you have 2, 3, 5 and 7, with two digits you only have 11. With three digits you have many palindromic primes, like 101 (the smallest) and 919 (the largest). With any even number of digits $d \geq 4$ you have no palindromic prime.

Paulo. Is this easy to prove?

P.P. Quite easy. I give the proof for palindromic primes with 6 digits. It is the same proof for any even number $d \neq 2$ of digits. Let N be a palindromic number with 6 digits. This means that (in the base 10 notation, the usual notation) $N \underset{(10)}{=} abccba$ where a, b, c are digits among $0, 1, 2, \ldots, 9$. So $N = a + b \times 10 + c + c \times 10^2 + c \times 10^3 + b \times 10^4 + a \times 10^5 = a(1 + 10^5) + b(10 + 10^4) + c(10^2 + 10^3)$. But $10^5 + 1 = (10 + 1)(10^4 - 10^3 + 10^2 - 10 + 1)$ (I did this calculation several times), so $10^5 + 1$ is divisible by 11, also $10^4 + 10 = 10(10^3 + 1) = 10(10 + 1)(10^2 - 10 + 1)$, so $10^4 + 10$ is divisible by 11, and finally, $10^3 + 10^2 = 10^2(10 + 1)$ is divisible by 11. It follows that N is divisible by 11.

Paulo. I could have done it by myself, so easy it was, Papa Paulo. Next time you should force us to do such easy proofs.

P.P. There is something that has never been proved: For every odd $d > 3$ there exists a palindromic prime with exactly d digits.

Eric. What is your feeling about this statement? True or false?

P.P. Feeling is the right word. How do I nurture this feeling? I tell you there are so many prime numbers that if I do not find any good reason right away to say that it is "false", then I say it is "true".

Eric. I am flabbergasted with your way of thinking.

P.P. I said "true". Prove me wrong. In the meantime I tell you what follows if the assertion is indeed true. Let $k > 1$ and $d \geq 1$ be given numbers, with d odd. I may find a sequence of palindromic primes as follows. P_1 has d digits, P_2 has P_1 digits, P_3 has P_2 digits and so on, P_k has P_{k-1} digits, all the numbers P_i are palindromic primes.

Eric. That is incredible. Can you show me an example?

P.P. Yes but only for $k = 3$. The following notation is practical:
$$(23)_4 = 23\,23\,23\,23$$
$$(1)_6 = 1\,1\,1\,1\,1\,1$$
and
$$5(2)_3(3)_4\,2 = 5\,222\,3333\,2.$$

For the example I take
$$d = 1,\ k = 3,\ P_1 = 5,\ P_2 = 3\,5\,3\,5\,3$$
and
$$P_3 = 1(0)_{17672}\,24042\,(0)_{17672}\,1.$$

An example with $k = 4$ is presently out of reach.

Paulo. Amazing. Is P_3 the largest known palindromic prime?

P.P. I know one which is bigger. Not long ago, it was the largest known palindromic prime. It is $(9)_{52140}\,8\,(9)_{52140}$.

Paulo. Are palindromic primes important, or are they no more than primes with digits following a nice pattern?

P.P. The patterns for the digits are present in the usual decimal basis writing. Look at the number written 11 in decimal basis. In basis 2, $11 = 1011_{(2)}$ which is no longer palindromic. Properties of digits in base 10 do not confer any importance to the number.

P.P. Now I will give a list of primes with interesting digits. Get yourself ready for a nice list:

(a) The largest known prime all of whose digits are prime numbers is
$$(72\,32\,32\,52\,32\,32\,72\,32\,52\,52)_{156} + 1.$$
It has 3120 digits.

(b) The largest known prime whose digits are 0 or 1 is
$$1\,(0)_{15397}\,1110111\,(0)_{15397}\,1.$$
It has 30803 digits.

(c) The largest known primes with all digits equal to 9, except the initial digit, are
$$1\,(9)_{55347}$$
$$2\,(9)_{49314}$$
$$4\,(9)_{21456}$$
$$5\,(9)_{34936}$$

$7\,(9)_{49808}$

$8\,(9)_{48051}.$

(d) The largest known prime all of whose digits are odd is

$$1\,(9)_{55348}.$$

(e) The largest known prime with the largest number of digits equal to 0 is

$$105994 \times 10^{105994} + 1.$$

(f) The most exotic prime is

$(1)_{1000}\,(2)_{1000}\,(3)_{1000}\,(4)_{1000}\,(5)_{1000}\,(6)_{1000}\,(7)_{1000}\,(8)_{1000}$
$(9)_{1000}\,(0)_{6645}\,1.$

It has 15646 digits.

And last (and least)

(g) The smallest prime with 1000 digits is

$$1\,(0)_{998}\,7.$$

Paulo. What a display of curiosities!

Eric. Now I understand what you said before, Papa Paulo. If nothing immediately detected prevents that a prime exists with a certain property, then such a prime actually exists, and most likely, infinitely many of them. But, except in a few cases, no one is able to prove it.

19

Secret Messages

P.P. You are in good company. Have you ever heard the name Leonardo da Vinci?

Eric. Yes, of course. He painted the famous lady with an enigmatic smile.

P.P. He was not only one of the very great painters, but also a sculptor, an architect, an engineer and a very imaginative inventor.

Paulo. I heard that tens of thousands of people come from all over the world to visit the Louvre Museum just to see that smile and contemplate its meaning. But why are you talking about smiles? You promised secret messages.

P.P. Leonardo created many inventions, made detailed drawings and wrote how they should be built and function. To prevent anyone else from stealing his ideas, he wrote the text backwards as if it were a mirror image. In his handwriting, the texts would not be easy to read by any other person.

Eric. I would like to see these texts. Tomorrow, I'll go to the Art Library of the nearby university and examine Leonardo's notebooks.

Eric took over:

Eric. This is pretty easy to decipher. It was fine at the 16th century but no more today.

Paulo. I like stories of war and war messages.

P.P. The stories may be good, but wars are not. Many people are killed, many people suffer.

Eric. I heard about pigeons carrying written messages tucked into their bodies. It could be an order for troops to attack. It used coded words like this famous one which was important in a battle: "Eat breakfast at 5 a.m. with 15 pieces of bread." If the message was intercepted by the other side, they would think: "The enemies are really hungry in the morning." But the general who received the message knew its meaning: "Attack the enemy at 5 a.m. with 15 battalions of cavalry."

Paulo. The enemies were no fools. No pigeon would travel with such instructions for breakfast. It was easy to guess that the message was an order to attack in the morning. I don't think that this was a safe system to send secret messages.

Eric. During the last war they used another method which consisted of scrambling the letters of the alphabet.

Paulo. How does it work?

Eric. You replace the letters A, B, C, D, E, ... by other letters, say, by F, E, G, A, B, ... So the word "ace" is now printed "fgb." Only you and your correspondent have the key. There are so many ways of scrambling letters. To discover the key and decipher the scrambled text appears to be a hard challenge.

Paulo. 26 letters can be scrambled in 26! ways. Remember, 3 letters can be scrambled in 3! = 6 ways ABC, ACB, BAC, BCA, CAB and CBA. 4 letters may be scrambled in 4! = 24 ways, 5 letters in 5! = 120 ways. And ... I don't want to calculate 26! — it is very, very big.

Eric. These messages must have been impossible to decipher without the key.

P.P. Not so. There is a very clever method to decipher such messages; it works for texts which are not too short. You count how many times each letter appears in the encoded message. You have to know the frequency of letters in any text written in each of the major languages. Say, you assume that the text is in English. The relative frequency of the various letters in an English text is known, as you can see from the table.

To decode the scrambled text, just consider the frequency of any letters appearing in the encoded text, then compare with the table of frequency of letters in the English text. This establishes a correspondence between the letters of the encoded text and the letters of the text as it will be decoded.

By letter		By frequency	
Letter	Frequency	Letter	Frequency
a	0.08167	e	0.12702
b	0.01492	t	0.09056
c	0.02782	a	0.08167
d	0.04253	o	0.07507
e	0.12702	i	0.06966
f	0.0228	n	0.06749
g	0.02015	s	0.06327
h	0.06094	h	0.06094
i	0.06966	r	0.05987
j	0.00153	d	0.04253
k	0.00772	l	0.04025
l	0.04025	c	0.02782
m	0.02406	u	0.02758
n	0.06749	m	0.02406
o	0.07507	w	0.02360
p	0.01929	f	0.02228
q	0.00095	g	0.02015
r	0.05987	y	0.01974
s	0.06237	p	0.01929
t	0.09056	b	0.01492
u	0.02758	v	0.00978
v	0.00978	k	0.00772
w	0.02360	j	0.00153
x	0.00150	x	0.00150
y	0.01974	q	0.00095
z	0.00074	z	0.00074

Paulo. I will be very happy to decode a scrambled text message, but in English, please. Papa Paulo, can you give me a message that is not too short? I want to find the key.

P.P. Give me a few minutes, while I prepare the key and write the message with my computer.

Exactly seven minutes later, Papa Paulo proposed this message and, to give some trouble, there was no separation between the words (how mean!).

Scrambled text:

JEXHGRIQXBRYNTCJTDFATCETDFHGTBXDABIXGCYGRBVRBHRC
TJXDABIXGCMDNRYGXCRQZTDFJEXQMJJXGTDJRJEXTGHGTBXY
MVJRGCTCLDRKDJRIXRDXRYJEXBRCJTBHRGJMDJMDNACXYAQT
DMGTJEBXJTVJEXNTFDTJPRYJEXCVTXDVXTJCXQYCXXBCJRG
XOATGXJEMJXZXGPHRCCTIQXBXMDCIXXWHQRGXNYRGJEXCRQ
AJTRDRYMHGRIQXBCRXQXFMDJMDNCRVXQXIGMJXN

Paulo. I think that it will be hard to decode the message, but I will try.

Eric. Papa Paulo, your explanation was crystal clear. So, messages encoded as you described are easy to decode because they are lacking security.

Paulo. I understand that there was a pressing need for security in secrecy.

P.P. Yes, let me tell you an incredibly better method.

Public Key Cryptography

P.P. Owing to the proliferation of means of communication and the need to send messages — like bank transfers, instructions for buying stocks, secret diplomatic information, as, for example, reports of spying activities — it has become very desirable to develop a safe method of coding messages. In the past, keys have been kept secret, known only to the parties sending and receiving the messages, but it has often been possible to study the intercepted messages and crack the code. As I showed, in simpler cases it would be enough to study the frequency of symbols in the message. In war situations, this had disastrous consequences.

Great progress in cryptography came with the advent of public key crypto-systems. The main characteristics of the system are its simplicity, the public key, and the extreme difficulty in cracking it. The idea was proposed in 1976 by Diffie and Helman, and the effective implementation was proposed in 1978 by Rivest, Shamir and Adleman. This crypto-system is therefore called the RSA-system. I will describe it:

Each letter or sign, including blank space, corresponds to a 3-digit number. In the *American Standard Code for Information Interchange* (ASCII), this correspondence is the following:

—	A	B	C	D	E	F	G	H
032	065	066	067	068	069	070	071	072
I	J	K	L	M	N	O	P	Q
073	074	075	076	077	078	079	080	081
R	S	T	U	V	W	X	Y	Z
082	083	084	085	086	087	088	089	090
COMMA		PERIOD						
044		046						

Each letter or sign of the message is replaced by its corresponding 3-digit number, giving rise to a number M, which represents the message.

For example, the sentence
$$\text{I LOVE YOU JENNIE}$$
becomes
$$073032076079086069032089079085032074069078078073069$$

Each user A of the system has a key. The key consists of a pair of positive integers (n_A, s_A). This key is public, just like a telephone number appears in a directory. The number n_A is the product of two odd prime numbers p_A and q_A, that is, $n_A = p_A q_A$. The numbers p_A and q_A are only known to the user A — to no one else. And s_A has to be coprime to $p_A - 1$ and to $q_A - 1$.

Now I will explain how user A sends a secret message to user B.

Step 1: The English text is transformed into a sequence of digits which constitute the number M, in the way I have illustrated for the sentence "I love you Jennie".

Step 2: Breaking M into blocks: A looks at the key (n_B, s_B) of B, to whom he wants to send the message M. If $M > n_B$ then M is broken into several blocks $M_1, M_2, \ldots, M_r$. Each block is a number less than n_B, consisting of a succession of integers with three digits, representing letters. For example, if $n_B = 156287$ and $s_B = 181$, $M = 088077069088032065$ then $M = M_1 M_2 M_3$ where the blocks are $M_1 = 88077$, $M_2 = 69088$, $M_3 = 32065$.

Step 3: Preparing the blocks: If $\gcd(M_i, n_B) = 1$ let $\overline{M}_i = M_i$. If $\gcd(M_i, n_B) \neq 1$ then either p_B or q_B divides M_i, so $1 < M_i$, let $\overline{M}_i = M_i - 1 \geq 1$.

Step 4: Encoding the prepared blocks: $\overline{M}_i$. Let M'_i be such that $1 \leq M'_i < n_B$ and $M'_i \equiv \overline{M}_i^{s_B} \pmod{n_B}$. Let M' be the number with blocks $M'_1, M'_2, \ldots, M'_r$. M' is the encoded message received by B.

Step 5: Decoding the messages M'_i: From $\gcd(s_B, (p_B-1)(q_B-1)) = 1$ and $\varphi(n_B) = (p_B-1)(q_B-1)$ there exists t_B such that $1 \leq t_B < \varphi(n_B)$ and $t_B s_B \equiv 1 \pmod{\varphi(n_B)}$, that is $t_B s_B = 1 + k_B \varphi(n_B)$, where k_B is an integer. Then $M'^{t_B}_i \equiv \overline{M}_i^{s_B t_B} \equiv \overline{M}_i^{1+k_B \varphi(n_B)} \equiv \overline{M}_i \pmod{n_B}$, because $\gcd(\overline{M}_i, n_B) = 1$ so $\overline{M}_i^{k_B \varphi(n_B)} \equiv 1 \pmod{n_B}$ by Euler's theorem. Then either $\overline{M}_i$ or $\overline{M}_i + 1$ can be translated into a meaningful sequence of English letters. This being done for each $i = 1, \ldots, r$, leads to the decoded message M_i sent by A.

Paulo. The process is very simple.

Eric. And very clever.

Eric. You did not discuss the security of the method.

P.P. Look back at the steps. To decode the message received by B, all he reeds to know are the prime factors p_B and q_B of n_B. Then we get $\varphi(n_B) = (p_B-1)(q_B-1)$ and from this it is possible to calculate t_B and this will, in turn decode the message received by B.

Eric. So, for security purposes, n_B has to be hard to factor.

P.P. The thing to do is to define n_B as being the product of two equally big primes. Then it is usually hard to find the factors of n_B. Of course it all depends on sizes as well as on present day methods of factorization. The better these methods become, the bigger n_B has to be to resist the factorization procedure.

Eric. Has anyone found a factorization method which runs in polynomial time for generic integers?

P.P. Up to now this is the situation. So I can say, it is considerably longer to factor (generic) numbers than to test for primality. And this is the basis of the security for the RSA crypto-system.

Paulo. At present, what is the approximate size of the numbers n_B to offer sufficient security?

P.P. I am not sure anymore. It is evolving so fast. Certainly numbers n_B with 300 digits offer enough security.

Eric followed the discussion and seemed impatient, wanting to intervene.

Eric. Go back to the decoding process. We only need to know $\varphi(n_B)$. So you try all numbers h such that $1 \leq h < n_B$, you calculate $\gcd(h, n_B)$ and detected how many h's there are such that $\gcd(h, n_B) = 1$ — this count gives $\varphi(n_B)$. And that is what is important in my view. To calculate the gcd you only use Euclidean division algorithm and no factorization, which is much harder to perform.

P.P. That is a very sharp remark, Eric. But note, there are so many numbers h to investigate that there is no gain.

Eric wanted to say what seemed important. He sees all sorts of maneuvers that people could do, so he said:

Eric. Papa Paulo, one day I will become the director of a bank. Like the user B, I will receive messages which are encoded and after decoding, one may say: "Transfer US$100 000 from my account to the account of C in bank so-and-so. Signed, A." With such an amount of money, I want to be sure that the message is from A, not forged by C. Is there any way to know that a message comes with a guaranteed signature?

Paulo. I never thought about forged messages. This is an important security problem.

P.P. Yes, there is a way to send "signed messages".

Eric. Can you explain it to us? Is it difficult?

P.P. You be the judge. Here is how A sends a signed message to B, so that B is sure that it comes from A and therefore it is not forged. As before, it is enough to deal with messages M which are small, in our case, $M + 4 < \ell \subset \text{lcm}(n_A, n_B)$. If $n_A = p_A q_A$ and $n_B = p_B q_B$, then ℓ is divisible by at most four distinct primes. So one of the numbers M, $M+1$, $M+2$, $M+3$, $M+4$ is coprime to ℓ, otherwise there would exist one of the primes dividing ℓ and two of the five numbers M, $M+1$, $M+2$, $M+3$, $M+4$, which is impossible. Changing notation if needed, we may assume that $M < \ell$ and $\gcd(M, n_A) = 1$, $\gcd(M, n_B) = 1$. Moreover, $\gcd(M, n_A) = 1$ and $\gcd(M, n_B) = 1$.

$$M' \equiv \overline{L}^{s_A} \equiv L^{t_B s_A} \equiv \overline{M}^{s_B t_B s_A} \equiv M^{t_A s_B t_B s_A}$$
$$\equiv M^{(1+h\varphi(n_A))(1+k\varphi(n_B))}$$
$$\equiv M^{1+h\varphi(n_B)} \equiv M \pmod{\ell}.$$

Since $M' < \ell$ and $M < \ell$, then $M' = M$.

So B may read the message M, which was decoded using the key of A. Therefore B knows that message comes from A.

The following steps are needed.

Step 1: A calculates $\overline{M} < \ell$ such that $\overline{M} \equiv M^{t_A} \pmod{\ell}$.

Step 2: A uses the public key of B and calculates $L < \ell$, $L \equiv \overline{M}^{s_B} \pmod{\ell}$. The message L is sent to B.

Step 3: B uses his secret decoding method and obtains the message $\overline{L} < \ell$, such that $\overline{L} \equiv L^{k_B} \pmod{\ell}$.

Step 4: B uses the public key of A and obtains the message $M' < \ell$, $M' \equiv \overline{L}^{s_A} \pmod{\ell}$.

By definition, there exist integers h, k such that $s_A k_A = 1 + h\varphi(n_A)$ and $s_B t_B = 1 + k\varphi(N_B)$.

P.P. Electronic fraud is a very common crime involving millions and millions of dollars. These problems have forced specialists in cryptography to invent methods to recognize fraud. These methods are for the most part kept secret. I am not able to say more. Oh, yes, just one bit more. Cryptographers are at least as clever as those people who perform fraudulent operations. Crime does not pay. Sooner or later the bad guys will end up in prison.

Now, Eric demanded an encoded text to practice decoding. I had nothing ready, so I said:

P.P. Tomorrow morning you will see my text to decode.

The next morning, I gave them:

Message from A to B:

key of $B: n_B = 156287$, $s_B = 181$

123491 155914 128320 111475 64511 72506 21141 86164
151474 71596 36253 32006 146303 32461 11098 52772 102274
53595 71596 36253 69087 56547 14973 133631 41725 21141
28062 121367 126510 122000 33120 56547 154303 133631
121367 40150 71596 36253 113085 108419 25430 133631
80834 20213 71596 36253 69087 1003 50259 151474 133631
27323 74706 121367 126510 56547 154303 133631 113338
20213 76924 121367 126510 79033 39470 133631 57078 30779
93290 121367 126510 56547 154303 16276 155914 95179
28062 121367 126510 7464 130805 53457 102274 99054 71596
36253 16679 99054 5776 28062 50259 153075 121367 126510
76974 102274 3772 96642 46607 43449 155914 128320 111475
126510 155904 102274 114193 12481 64929 44476

20

New Numbers and Functions

P.P. The topic today will be the real numbers and functions.

New Numbers

Papa Paulo wanted to present the real numbers in a very simple and concrete way.

P.P. Numbers like $x = 35.01875328\ldots$ are called *real numbers*. 35 is the *integral part* of x, $.01875328\ldots$ is the *decimal part* of x.

The decimal part may be finite, like in $0.48 = 0.48000,\ldots$, or *periodic* like in $3.555\ldots$ (with period 5), or like $12.45131313\ldots$ (with period 13).

The numbers x with finite or periodic decimal part are rational numbers. For example

$$0.48 = \frac{48}{100} = \frac{12}{25},$$

$$2.333\ldots = 2 + \frac{3}{10} + \frac{3}{10^2} + \frac{3}{10^3} + \cdots$$

$$= 2 + \frac{3}{10}\left(1 + \frac{1}{10} + \frac{1}{10^2} + \cdots\right)$$

$$= 2 + \frac{3}{10} \times \frac{1}{1 - \frac{1}{10}} = 2 + \frac{3}{10} \times \frac{10}{9}$$

$$= 2 + \frac{1}{3} = \frac{7}{3}.$$

In the same way, $0.999\ldots = 1$.

If you remember, $\sqrt{2}$ is not a rational number, so the decimal part of $\sqrt{2}$ is infinite and not periodic.

A real number which is not a rational number is called an *irrational* number.

Eric. From your examples, it is obvious that if the real number x is not a rational number, then it can be approximated as close as one likes by a rational number. I do like you, Papa Paulo, I give an example: $1.413587\ldots - 1.4135 = 0.000087\ldots = \frac{87}{1000000}\ldots$.

Functions

P.P. Imagine that you have a set D of real numbers and a rule f which associates to every real number x of D a real number, which we denote by $f(x)$.

We call f a *function* and D is the *domain* of f. The real numbers $f(x)$ are the *values* of f and the set of all values is the *range* of f.

Paulo. Papa Paulo, you should give some examples of functions.

P.P. The functions $f(x) = x + 3$, $f(x) = x^2$ have domain and range equal to the set of all real numbers. The function $f(x) = \frac{1}{x}$ has domain equal to the set of all non-zero real numbers and range equal to the set of all real numbers. I give one more example, the function $f(x) = \sqrt{x}$, with domain and range equal to the set of all $x \geq 0$.

The *graph* of a function f consists of all the parts in the plane with coordinates (x, y), where $y = f(x)$. Plotting the graph helps to visualize the function on the set of all real numbers.

Papa Paulo continued:

P.P. It is very important to describe how the value of the function changes when the real number x changes, or varies. For this reason we like to call x the *variable*.

Paulo. Papa Paulo, examples, examples!

P.P. When x approaches 2, then $x+3$ approaches 5 and x^2 approaches 4, while $\frac{1}{x}$ approaches $\frac{1}{2}$ and $\sqrt{2}$ approaches $\sqrt{2}$. We like to say that the value approached its *limit* and we write:

$$\lim_{x \to 2} x + 3 = 5, \quad \lim_{x \to 2} x^2 = 4,$$

$$\lim_{x \to 2} \frac{1}{x} = \frac{1}{2}, \quad \lim_{x \to 2} \sqrt{x} = \sqrt{2}.$$

Eric. Can we also consider the limit of $f(x)$ when x approaches infinity, that is, x grows to become arbitrarily large?

P.P. This is not only possible, but very important this time. I will give the definition of these limits as mathematicians do. The definitions express in a rigorous manner the intuitive sense of approximation to the limit. Here it goes.

Eric. I agree with you. You can perceive that $f(x)$ approaches a value. No problem.

We concentrate our attention on a function $f(x)$ defined for all real numbers $x > 0$. We say that the *limit* of $f(x)$ as x tends to (or approaches) infinity is equal infinity when for every natural number N (however large it may be) there is a natural number M (which depends on the given number N) such that if $x > M$ then $f(x) > N$. In this situation we write $\lim_{x \to \infty} f(x) = \infty$. In words, $f(x)$ becomes arbitrarily large when x is sufficiently large.

The functions $f(x) = x + 2$ and $f(x) = x^2$ are such that $\lim_{x \to \infty} f(x) = \infty$.

A companion definition is the following. Let A be a number. We say that A is the limit of $f(x)$ when x tends to infinity, when for every natural number $N > 0$ there is a natural number $M > 0$ (which depends on N) such that if $x > M$ then $A - \frac{1}{N} < f(x) < A + \frac{1}{N}$. Now we write $\lim_{x \to \infty} f(x) = A$. In words, $f(x)$ is arbitrarily close to A when x is sufficiently large. If $f(x) = \frac{1}{x}$ then $\lim_{x \to \infty} f(x) = 0$.

Paulo. Papa Paulo, you only considered limits when x tends to infinity. Is this all one needs?

P.P. Actually we need the definition of limit when x approaches a number, say, like 1, from the right (or from the left, or from right or left). But this definition is similar and you may spell it out by yourself. Think a little.

Paulo. Papa Paulo, you just gave the definition of the limit of a function at a point. But you did not say if every function has a limit at every point sx. Tell me about this.

P.P. A function with erratic behavior, like one with oscillating values, may not have a limit at certain points. One learns at university, in the study of functions, limits in much detail. But here I stick to simple situations.

Now Eric brought back limits for functions with domain the positive integers.

Eric. The definitions of limits, I suppose, apply also for sequences. Can you phrase it very explicitly?

P.P. No problem. Let $s_1, s_2, s_3, \ldots$ be a sequence of numbers. We say that the sequence of s_n's has limit infinity if for every integer $N > 0$ there exists an index $n_0 \geq 1$ such that if $n > n_0$, then $s_n > N$. We write $\lim_{n \to \infty} s_n = \infty$.

Let A be a number. We say that the sequence of s_n's has limit A when for every integer $N > 0$ there is a natural number n_0 such that if $n > n_0$ then $A - \frac{1}{N} < s_n < A + \frac{1}{N}$. We write $\lim_{n \to \infty} s_n = A$.

I add that not every sequence has a limit when n tends to infinity. I am surprised nobody asked me, but I will say it anyway. If $f(x)$ is a function, it cannot have two different limits when x tends to a value, infinite or finite. The same holds of course for any sequence, it cannot have two different limits when n tends to infinity.

Paulo. I know it already; the values of the function or the terms of the sequence could not be arbitrarily large and at the same time arbitrarily close to a number, nor could they be arbitrarily close to two different numbers.

Eric. Everything you said is clear. I am glad to learn the concepts of limit, I know that it may be difficult to compute certain limits.

I know also that if you will refer to standard properties of limits, I will accept them without pain. But, I think, you did not do all this to deal with functions as simple as $f(x) = x+2$, $f(x) = x^2$, $f(x) = \frac{1}{x}$.

P.P. You are right and I shall have to consider the logarithmic function and the exponential function. This has to be preceded by the evaluation of areas.

The Evaluation of Areas

P.P. I begin with something very simple, the area of a rectangle with base measuring $x > 0$ and height measuring $y > 0$.

Paulo. I know, it is xy.

P.P. If I have several rectangles, each one with base measuring x and heights $y_1, y_2, \ldots, y_n$, (and they are not overlapping) the sum of the areas is $xy_1 + xy_2 + \cdots + xy_n = x(y_1 + y_2 + \cdots + y_n)$. OK, but I want to measure more complicated areas. Let $f(x)$ be a function with values $f(x) \geq 0$, let x' and x'' be in the domain of $f(x)$ with $x' < x''$. We assume that if $x' < x < x''$, then x is in the domain of the function. Consider all the points (x, y) with $x' \leq x \leq x''$ and $0 \leq y \leq f(x)$. This set of points has an area, which I want to evaluate.

Eric. It may be hard if $f(x)$ is erratic. It is even conceivable...

Eric paused for a while, then said:

Eric. How erratic can erratic be? Maybe there is no unquestionable way of measuring the area.

P.P. What you said is a possibility to worry about. But I am only interested in the areas when $f(x)$ is continuous, the values do not jump.

Paulo. I wonder how you can measure the area. Usually the graph of $f(x)$ is a curve, not made up of horizontal sides of rectangles.

P.P. The idea is as simple as it is old. Say $x'' - x' = d$, let N be a positive integer.

Let $[x', x' + \frac{d}{N}]$ be the segment of the x-line with left extremity x' and right extremity $x' + \frac{d}{N}$. Consider the analogous segments $[x' + \frac{d}{N}, x' + \frac{2d}{N}], [x' + \frac{2d}{N}, x' + \frac{3d}{N}], \ldots, [x' + \frac{(N-1)d}{N}, x'']$.
We take these segments as bases of rectangles. Each rectangle has to be below the graph of $f(x)$ and to be as high as possible. As I said, it is easy to find the sum of the areas of these rectangles, once I know their heights. This sum of areas of rectangles depends not only on the function $f(x)$, but also on the integer N. I call it A_N. So for every positive integer N, I have a real number A_N which is less or equal to the area A to be evaluated. Thus I obtain a sequence of real numbers $A_1, A_2, \ldots, A_n, \ldots$ and if $N < N'$ then $A_N \leq A_{N'}$. By definition, the area A is $A = \lim_{N \to \infty} A_N$. I want to add something. It is intuitive and it may be proved that the limit in question exists and it is finite.

Eric. Papa Paulo, what are you going to do with areas?

P.P. One thing I am not going to do is to bore you with simple and intuitive properties of the area. But I wish to introduce the notation that is commonly used. It is a peculiar notation. As I said, the area is approximated by the sums of areas of rectangles. "Sum" suggests the letter S. As the sums to approximate the area uses rectangles with smaller and smaller basis and a limit is taken, the S is replaced by an elongated S, which looks like $\int$. It is an elegant symbol, which I like very much to look at.

Eric. But the name *elongated S* is not nice.

P.P. We call $\int$ the *integration symbol*. It is used to measure the area above the segment $[x', x'']$ and below the graph of $f(x)$, so we write $\int_{x'}^{x''} f(t)\, dt$ and we read: the *integral* of f from x' to x''.

Eric. Why this dt? It looks superfluous to me.

P.P. At first sight it is. The presence is to remind us that it is a limit of rectangles with smaller and smaller bases.

Paulo. Notations are notations. Probably in China, it would be different. But what counts is the rigor and clarity in the definition. You make it clear, Papa Paulo.

P.P. I accept your compliment with pleasure. But I have to be honest with you. The idea of definition of the area is what I said, simple and natural. But there are technical problems when one wants to prove that the limit defining the areas exist — this has a lot to do with the behavior of the function $f(x)$. Also, I skipped any explanation about the symbol dt. What I said would require further considerations to become meaningful. The keyword is *infinitesimal*. The whole matter of limits, integrals and (what I avoided saying) derivatives belongs to the branch of mathematics called *analysis*. To study primes, analysis will be needed; this has forced me to these inroads.

Eric. You explain all this so well, I wish that one day that you conduct discussions on the theme "Explaining functions to a million".

Eric's suggestion was ignored by Papa Paulo, who continued.

P.P. I declare that I will define, one after the other, the logarithmic function, the exponential function and the power function.

Three Functions

P.P. Let $f(x) = \frac{1}{x}$, defined for all $x > 0$. If $x > 1$ we consider, as I just did, the area above the segment $[1.x]$ and below the graph of $f(x) = \frac{1}{x}$. This area is called the *logarithm* of x. I define also $\log 1 = 0$, and if $0 < x < 1$, the definitions $\log x = -\log \frac{1}{x}$. Note that now $\frac{1}{x} > 1$, so $\log \frac{1}{x}$ was defined earlier.

Using the integral symbol, if $x > 1$:

$$\log x = \int_1^x \frac{dt}{t}$$

(integral of $f(t) = \frac{1}{t}$ from 1 to x).

Eric. This is the first time in my whole life (perhaps short, but whole) that I see a function defined by the measurement of areas. Mathematicians, you do not lack inventiveness.

P.P. Actually there are other ways of defining the logarithmic function. I could not use the other ways because we have not discussed the necessary background. Now I want to list the main properties of the logarithmic function.

(a) The logarithmic function is an increasing function: if $0 < x' < x$, then $\log x' < \log x$. In particular, if $0 < x' < 1$, then $\log x' < 0$.
(b) $\lim_{x \to \infty} \log x = \infty$.
(c) the domain of the function $\log x$ is the set of all $x > 0$.
(d) the range of the function $\log x$ is the set of all real numbers.
(e) for each real number y there exists one and only one real number x such that $\log x = y$.

Paulo. Which number has logarithm equal to 1?

P.P. This number is called e. The value of e is $2.7183\ldots$

Papa Paulo continued:

P.P. The most remarkable property of the logarithmic function allows us to replace the computation of a product by a sum:

(f) $\log(xx') = \log x + \log x'$.

Paulo. This is a good list of properties. Are you going to prove them?

P.P. No, the proofs of some of these properties, even though not difficult, require facts about the theory of integrals which we have not discussed.

And Papa Paulo continued unabated, presenting Eric and Paulo with another important function.

P.P. The *exponential function* is defined in the following way. If y is a real number, let x be the unique positive real number such that $\log x = y$. By definition, $\exp(y) = x$.

So I can also write $\exp(\log x) = x$ and $\log(\exp(y)) = y$.

I list some properties of the exponential function. They follow easily from properties of the logarithmic function

(a) $\exp(0) = 1, \exp(1) = e$.
(b) the domain of the exponential function is the set of all real numbers.

Paulo. Yes, because it is the range of the logarithmic function. And you are going to state the property.

(c) The range of the exponential function is the set of all positive real numbers.

Reason: it is the domain of the logarithmic function.

P.P. We also have these properties:

(d) The exponential function is an increasing function, that is, if $x < x'$, then $\exp(x) < \exp(x')$.
(e) $\lim_{x \to \infty} \exp(x) = \infty$.
(f) $\exp(x + x') = \exp(x) \cdot \exp(x')$.

P.P. I still need to discuss *powers*. Let abc be any positive real number. If n is a positive integer, we know that a^n = product of a with itself n times. We also have $a^0 = 1$. And we define $a^{-n} = \frac{1}{a^n}$ for all positive integers n. Just like for square roots of positive real numbers, there exists a unique real number $b > 0$ such that $b^n = a$. We say that $a^{\frac{1}{n}} = b$ and $a^{\frac{m}{n}} = b^m$.

Paulo. Up to now, it is easy and understandable. What do you want to do now?

P.P. I want to define a^x for any real number x. This is the good definition:
$$a^x = \exp(x \log a).$$
It makes sense and agrees with the preceding definitions when $x = \frac{m}{n}$.

Eric. Let me try to prove it. If $x = \frac{m}{n}$, then
$$\exp\left(x \log a\right) = \exp\left(\frac{m}{n} \log a\right)$$
$$= \exp\left(\log a^{\frac{m}{n}}\right) = a^{\frac{m}{n}}.$$

Easy.

P.P. If x, x' are real numbers and $x < x'$, then $a^x < a^{x'}$. Also $a^{x+x'} = a^x \, a^{x'}$.

Paulo. What happens when you take $a = e$, that number with $\log e = 1$?

P.P. Easy:
$$e^x = \exp(x \log e) = \exp(x).$$

Eric. How nice, so we may write e^x instead of $\exp(x)$.

Eric. Any more functions for our pleasure?

P.P. I think we've had enough for today. It was a good introduction to real numbers, functions, integrals, and the special functions which we shall need.

21

Princeps Gauss

P.P. Already as a boy, Gauss loved numbers.

Eric. Then he must have loved the most important numbers — OUR prime numbers.

P.P. Yes, he probably spent days looking at, and nights thinking about, prime numbers. He studied his tables of primes, trying to discover patterns.

Paulo. I would also do the same but (*with a begging tone*) I do not have a table of primes.

P.P. My third eye had warned me that you, Paulo, would be asking for one, so I brought one with me and I am giving this table of primes to you both. It goes only up to 10 000. You may begin the search of patterns.

Primes up to 10 000

2	3	5	7	11	13	17	19	23	29
31	37	41	43	47	53	59	61	67	71
73	79	83	89	97	101	103	107	109	113
127	131	137	139	149	151	157	163	167	173
179	181	191	193	197	199	211	223	227	229
233	239	241	251	257	263	269	271	277	281
283	293	307	311	313	317	331	337	347	349
353	359	367	373	379	383	389	397	401	409
419	421	431	433	439	443	449	457	461	463

467	479	487	491	499	503	509	521	523	541
547	557	563	569	571	577	587	593	599	601
607	613	617	619	631	641	643	647	653	659
661	673	677	683	691	701	709	719	727	733
739	743	751	757	761	769	773	787	797	809
811	821	823	827	829	839	853	857	859	863
877	881	883	887	907	911	919	929	937	941
947	953	967	971	977	983	991	997	1009	1013
1019	1021	1031	1033	1039	1049	1051	1061	1063	1069
1087	1091	1093	1097	1103	1109	1117	1123	1129	1151
1153	1163	1171	1181	1187	1193	1201	1213	1217	1223
1229	1231	1237	1249	1259	1277	1279	1283	1289	1291
1297	1301	1303	1307	1319	1321	1327	1361	1367	1373
1381	1399	1409	1423	1427	1429	1433	1439	1447	1451
1453	1459	1471	1481	1483	1487	1489	1493	1499	1511
1597	1601	1607	1609	1613	1619	1621	1627	1637	1657
1663	1667	1669	1693	1697	1699	1709	1721	1723	1733
1741	1747	1753	1759	1777	1783	1787	1789	1801	1811
1823	1831	1847	1861	1867	1871	1873	1877	1879	1889
1901	1907	1913	1931	1933	1949	1951	1973	1979	1987
1993	1997	1999	2003	2011	2017	2027	2029	2039	2053
2063	2069	2081	2083	2087	2089	2099	2111	2113	2229
2131	2137	2141	2143	2153	2161	2179	2203	2207	2213
2221	2237	2239	2243	2251	2267	2269	2273	2281	2287
2293	2297	2309	2311	2333	2339	2341	2347	2351	2357
2371	2377	2381	2383	2389	2393	2399	2411	2417	2423
2437	2441	2447	2459	2467	2473	2477	2503	2521	2531
2539	2543	2549	2551	2557	2579	2591	2593	2609	2617
2621	2633	2647	2657	2659	2663	2671	2677	2683	2687
2689	2693	2699	2707	2711	2713	2719	2729	2733	2741
2749	2753	2767	2777	2789	2791	2797	2801	2803	2819
2833	2837	2843	2851	2857	2861	2879	2887	2897	2903
2909	2917	2927	2939	2953	2957	2963	2969	2971	2999
3001	3011	3019	3023	3037	3041	3049	3061	3067	3079
3083	3089	3109	3119	3121	3137	3163	3167	3169	3181
3187	3191	3203	3209	3217	3221	3229	3251	3253	3257
3259	3271	3299	3301	3307	3313	3319	3323	3329	3331
3343	3347	3359	3361	3371	3373	3389	3391	3407	3413
3433	3449	3457	3461	3463	3467	3469	3491	3499	3511
3517	3527	3529	3533	3539	3541	3547	3557	3559	3571
3581	3583	3593	3607	3613	3617	3623	3631	3637	3643
3659	3671	3673	3677	3691	3697	3701	3709	3719	3727
3733	3739	3761	3767	3769	3779	3793	3797	3803	3821
3823	3833	3847	3851	3853	3863	3877	3881	3889	3907
3911	3917	3919	3923	3929	3931	3943	3947	3967	3989

4001	4003	4007	4013	4019	4021	4027	4049	4051	4057
4073	4079	4091	4093	4099	4111	4127	4129	4133	4139
4153	4157	4159	4177	4201	4211	4217	4219	4229	4231
4241	4243	4253	4259	4261	4271	4273	4283	4289	4297
4327	4337	4339	4349	4357	4363	4373	4391	4397	4409
4421	4423	4441	4447	4451	4457	4463	4481	4483	4493
4507	4513	4517	4519	4523	4547	4549	4561	4567	4583
4591	4597	4603	4621	4637	4639	4643	4649	4651	4657
4663	4673	4679	4691	4703	4721	4723	4729	4733	4751
4759	4783	4787	4789	4793	4799	4801	4813	4817	4831
4861	4871	4877	4889	4903	4909	4919	4931	4933	4937
4943	4951	4957	4967	4969	4973	4987	4993	4999	5003
5009	5011	5021	5023	5039	5051	5059	5077	5081	5087
5099	5101	5107	5113	5119	5147	5153	5167	5171	5179
5189	5197	5209	5227	5231	5233	5237	5261	5273	5279
5281	5297	5303	5309	5323	5333	5347	5351	5381	5387
5393	5399	5407	5413	5417	5419	5431	5437	5441	5443
5449	5471	5477	5479	5483	5501	5503	5507	5519	5521
5527	5531	5557	5563	5569	5573	5581	5591	5623	5639
5641	5647	5651	5653	5657	5659	5669	5683	5689	5693
5701	5711	5717	5737	5741	5743	5749	5779	5783	5791
5801	5807	5813	5821	5827	5839	5843	5849	5851	5857
5861	5867	5869	5879	5881	5897	5903	5923	5927	5939
5953	5981	5989	6007	6011	6029	6037	6043	6047	6053
6067	6073	6079	6089	6091	6101	6113	6121	6131	6133
6143	6151	6163	6173	6197	6199	6203	6211	6217	6221
6229	6247	6257	6263	6269	6271	6277	6287	6299	6301
6311	6317	6323	6329	6337	6343	6353	6359	6361	6367
6373	6379	6389	6397	6421	6427	6449	6451	6469	6473
6481	6491	6521	6529	6547	6551	6553	6563	6569	6571
6577	6581	6599	6607	6619	6637	6653	6659	6661	6673
6679	6689	6691	6701	6703	6709	6719	6733	6737	6761
6763	6779	6781	6791	6793	6803	6823	6827	6829	6833
6841	6857	6863	6869	6871	6883	6899	6907	6911	6917
6947	6949	6959	6961	6967	6971	6977	6983	6991	6997
7001	7013	7019	7027	7039	7043	7057	7069	7079	7103
7109	7121	7127	7129	7151	7159	7177	7187	7193	7207
7211	7213	7219	7229	7237	7243	7247	7253	7283	7297
7307	7309	7321	7331	7333	7349	7351	7369	7393	7411
7417	7433	7451	7457	7459	7477	7481	7487	7489	7499
7507	7517	7523	7529	7537	7541	7547	7549	7559	7561
7573	7577	7583	7589	7591	7603	7607	7621	7639	7643
7649	7669	7673	7681	7687	7691	7699	7703	7717	7723
7727	7741	7753	7757	7759	7789	7793	7817	7823	7829
7841	7853	7867	7873	7877	7879	7883	7901	7907	7919

7927	7933	7937	7949	7951	7963	7993	8009	8011	8017
8039	8053	8059	8069	8081	8087	8089	8093	8101	8111
8117	8123	8147	8161	8167	8171	8179	8191	8209	8219
8221	8231	8233	8237	8243	8263	8269	8273	8287	8291
8293	8297	8311	8317	8329	8353	8363	8369	8377	8387
8389	8419	8423	8429	8431	8443	8447	8461	8467	8501
8513	8521	8527	8537	8539	8543	8563	8573	8581	8597
8599	8609	8623	8627	8629	8641	8647	8663	8669	8677
8681	8689	8693	8699	8707	8713	8719	8731	8737	8741
8747	8753	8761	8779	8783	8803	8807	8819	8821	8831
8837	8839	8849	8861	8863	8867	8887	8893	8923	8929
8933	8941	8951	8963	8869	8871	8999	9001	9007	9011
9013	9029	9041	9043	9049	9059	9067	9091	2103	9109
9127	9133	9137	9151	9157	9161	9173	9181	9187	9199
9203	9209	9221	9227	9239	9241	9257	9277	9281	9283
9293	9311	9319	9323	9337	9341	9343	9349	9371	9377
9391	9397	9403	9413	9419	9421	9431	9433	9437	9439
9461	9463	9467	9473	9479	9491	9497	9511	9521	9533
9539	9547	9551	9587	9601	9613	9619	9623	9629	9631
9643	9649	9661	9677	9679	9689	9697	9719	9721	9733
9739	9743	9749	9767	9769	9781	9787	9791	9803	9811
9817	9829	9833	9839	9851	9857	9859	9871	9883	9887
9901	9907	9923	9929	9931	9941	9949	9967	9973	

Eric. Can anyone, or could Gauss make discoveries just looking at the table? Looking superficially I can only see that apart from 2 there is no even number and apart from 5, no number ends with the digit 5. I also see from time to time two successive odd integers which are primes and also some unexpected big gaps between primes which are successive.

P.P. Of course, the unpredictable arrangement of primes did not escape Gauss's attention. He did not even try to explain it. Instead, Gauss had the idea of counting how many primes there are up to any given natural number N, like $N = 100, 200, \ldots, 1000, 1100, \ldots, 5000$, etc.

Paulo. Could he find a pattern with such a small table?

P.P. He worked with a much bigger table, going up to $3\,000\,000$.

Papa Paulo continued:

P.P. The largest table I have at home goes up to 100 000 and it is far bigger than the one I handed to you.

Eric. How much bigger?

P.P. This is exactly what Gauss wanted to understand. Here is what he thought and what he did. First, Gauss counted how many primes there are up to N. He called this number $\pi(N)$ (read *pi of N*, π is the Greek letter *pi*). He made a table of the numbers $\pi(N)$. For example

$$\pi(2) = 1 \qquad \pi(8) = 4$$
$$\pi(3) = 2 \qquad \pi(9) = 4$$
$$\pi(4) = 2 \qquad \pi(10) = 4$$
$$\pi(5) = 3 \qquad \pi(11) = 5$$
$$\pi(6) = 3 \qquad \pi(12) = 5$$
$$\pi(7) = 4 \qquad \pi(13) = 6$$

etc.

It is clear that $\pi(N) = \pi(N-1) + 1$ when N is a prime. To obtain a function defined for all real numbers $x > 1$, just define $\pi(x)$ to be the number of primes $p \leq x$. Of course, if x is a real number such that $N \leq x < N+1$, then $\pi(x) = \pi(N)$, because primes are natural numbers. We may plot the graph of the function $\pi(x)$. It consists of horizontal segments, not containing the right extremity, with upward jumps of 1 at every prime number N. You may easily draw the graph of the function.

Gauss looked at the graph of $\pi(x)$ and thought: Do I know any function which has this graph?

Paulo. Did Gauss know many functions?

P.P. He knew all the functions that anyone knew at his time. Like $f(x) = ax + b$ (with a, b real numbers, $a > 0$), or $f(x) = ax^2 + bx + c$ (with a, b, c real numbers, $a > 0$) and functions of similar type. He plotted the graphs of these functions. No jumps. Also for each $a > 0$, if he plotted the graph, as soon as x was sufficiently large, the value of $f(x)$ would surpass $\pi(x)$ and the difference would become bigger and bigger. So these functions would not fit well to approximate the values of $\pi(x)$. Gauss also tried functions which were the quotient of functions of the type considered above. No good. Gauss knew also the trigonometric functions, like $\sin x$ (sine of x), $\cos x$ (cosine of x), $\tan x$ (tangent of x) — you also know these functions from school. No luck with these either. Sine and cosine kept oscillating with values between -1 and 1 — bad for Gauss's purpose. Tangent had positive and negative values. No good. But Gauss also knew the exponential function and the logarithmic function. No jumps. As x grew the exponential function grows even faster than any function $f(x) = x^n$ (however big n may be). So the exponential function soon has values which are far too big with respect to $\pi(x)$, because $\pi(x) \leq x$.

Gauss also looked at the function $\log x$ which has no jumps and grows with x to become infinite. The opposite phenomenon happened. After a while he noted that $\pi(x) > \log x$.

Gauss was not desperate, he was not that kind of boy.

Paulo. You said "boy".

P.P. Yes, he was 15 at the time. As I said, he kept looking at the elusive function which would be fairly well comparable to $\pi(x)$.

After a short pause, Papa Paulo explained:

P.P. What Gauss wished to find, to observe the erratic behavior of the function $\pi(x)$, just around any number x. He felt that he could not say anything of interest for $\pi(x)$ when x belonged to short segments. Total mystery for him. Unpredictable. However, he was hopeful to find a function $f(x)$, with the same rate of growth as $\pi(x)$, when x grows to infinite and this is what was important, $f(x)$ should be *easily computable*. He knew, as I said, a number of computable functions and he searched an appropriate combination which would fit as tightly as possible the function $\pi(x)$. A bit like the Buenos Aires tailor.

To provoke, Eric said:

Eric. What do Buenos Aires and Gauss have to do with each other? Did Gauss dance the tango?

P.P. No, I don't think I have ever told one of my favorite stories. A customer, with an unusual body shape, asked a Buenos Aires tailor:

Unusual customer: Make for me a jacket from a fine fabric, I want to look good. When people see me from a distance, they will not realize that I have an unusual body. I am not like your everyday customer. But I do not want a loose poncho. I want a jacket that fits well. Tight but allowing for good movement. And (*he confided in a lower voice*), very elegant, so I will be noticed by women of good taste.

And to make himself important he concluded:

Unusual customer: I was warmly recommended by my good friend Wilfrid K.

This is my Buenos Aires tailor story and I tell you, that tailor did not relent until he satisfied his customer and above all, himself as well.

Papa Paulo continued:

P.P. You understand why I compared Gauss and the Buenos Aires tailor. Their tasks were similar. Gauss had to dress up $\pi(x)$ with a tightly fit and smooth function $f(x)$.

Paulo. Did Gauss succeed? All the other attempts you have described had failed. What did Gauss do?

P.P. Gauss thought, the function $h(x) = x$ grows too fast, the function $g(x) = \log x$ does not grow fast enough. He thought of dampening the growth of x by dividing it by $\log x$ and as far as Gauss could calculate with his tables, it seemed that $\pi(x)$ and $\frac{x}{\log x}$ had the same rate of growth. But Gauss also considered, besides $\frac{x}{\log x}$ another function, called the *logarithmic integral*, denoted by $\text{Li}(x)$ and defined by

$$\text{Li}(x) = \int_2^x \frac{dt}{\log t}$$

for $x \geq 2$. The logarithmic integral is the measure of the area below the graph of $\frac{1}{\log x}$, above the x-line, and between the vertical line containing the real number 2 and the vertical line containing the real number x, Gauss noted, which is not difficult, that $\frac{x}{\log x}$ and $\text{Li}(x)$ seemed to have the same rate of growth when x increases indefinitely. So $\text{Li}(x)$ would be a good function to compare with $\pi(x)$.

Eric. I can imagine that Gauss was getting excited. Indeed, $\frac{x}{\log x}$ and $\text{Li}(x)$ are computable and seemed to Gauss — you said seemed — to have the same rate of growth as $\pi(x)$.

P.P. Even at 15, Gauss was not naive. I want to explain carefully the tasks which were ahead for Gauss.

Task 1: To continue his observations calculating $\pi(x)$ for larger and larger values of x — for this he would need bigger tables of primes.

Task 2: To formulate exactly the theorem he thought to be true, concerning the growth rates of $\pi(x)$ and $\frac{x}{\log x}$, $\text{Li}(x)$.

Task 3: To evaluate how well $\frac{x}{\log x}$ and $\text{Li}(x)$ fit $\pi(x)$.

Task 4: Could he find a function $h(x)$ with the same rate of growth as $\pi(x)$ and which would approximate $\pi(x)$ even better than $\frac{x}{\log x}$ and $\text{Li}(x)$? — of course $h(x)$ should also be computable and smooth.

Eric. Already at 15 Gauss had worked for a lifetime.

The Tasks

P.P. All through his life Gauss did not cease to think about the distribution of prime numbers. Prince Gauss worked at the same time in all areas of mathematics, but he never forgot his beloved prime numbers.

For Task 1, as tables of primes were enlarged, Gauss could continue his observations. They confirmed his initial idea. But Gauss wanted more extensive calculations. Around 1840 a prodigious mental calculator named Zacharias Dase was performing the most extraordinary computations in his head, in a way that no one could explain. The number π of Ludolph — which I have mentioned earlier, was calculated by Dase, in two months, up to 200 decimal places, using an arc tangent formula proposed by Gauss. Dase performed other computational feats and gradually extended the table of primes up to 9 000 000. It was a pity that Gauss had died in the meantime.

Paulo. Neurologists still have to progress in the understanding of the brain, neurons, their connections, memory and whether intelligence is detectable somewhere in the brain.

Eric. Certainly children of intelligent parents are very often intelligent. Something in the genes.

Paulo. I read somewhere, but I forgot where, that only a limited fraction of our thinking resources is ever used.

Eric. And also memory, because you said "I forgot where."

Paulo. Did I?

Eric. You have not yet stated his theorem.

P.P. This was his Task 2, namely the exact formulation of what Gauss thought, better conjectured to be the Prime Number Theorem. To say that the functions $f(x)$ and $g(x)$ have the same rate of growth as x grows indefinitely (I can also say, as x tends to infinity) is no more no less than to say that the quotient $\frac{f(x)}{g(x)}$, or what is equivalent, the quotient $\frac{g(x)}{f(x)}$ has the limit 1. It is implicit that for all x, beyond some x_0, $f(x)$ and $g(x)$ are not equal to 0. In our present case, both $f(x)$ and $g(x)$ have positive values and grow to infinity as x grows to infinity.

The *Prime Number Theorem* stated, but not proved by Gauss, is the following:

Prime Number Theorem. $\lim \frac{\pi(x)}{\frac{x}{\log x}} = 1$, *which is just equivalent to* $\lim \frac{\pi(x)}{\operatorname{Li}(x)} = 1$ *(the limit is when x tends to infinity)*.

Of course, Gauss knew that the calculations he had made led to his proposed theorem but could not prove it. This is what was required to be proved:

For every natural number N (irrespective of how large N is chosen) there exists a natural number M (this number depends on N, the bigger N is, the bigger M has to be) such that the following inequalities are true: for every $x > M$:

$$1 - \frac{1}{N} < \frac{\pi(x)}{\frac{x}{\log x}} < 1 + \frac{1}{N}.$$

In words, it means that $\frac{\pi(x)}{\frac{x}{\log x}}$ is arbitrarily close to 1 for all x sufficiently large.

This is a statement which cannot be proved just by computations. The proof must appeal to a fairly deep knowledge of the distribution of primes.

Gauss — our Prince — our Number 1 mathematician, was not able to find this proof.

P.P. Now I want to discuss Task 3. We assume that the Prime Number Theorem had been proved, which was the hope of Gauss. From $\lim \frac{\pi(x)}{\frac{x}{\log x}} = 1$ we deduce that

$$\lim \left(\frac{\pi(x)}{\frac{x}{\log x}} - 1 \right) = 0,$$

that is

$$\lim \frac{\pi(x) - \frac{x}{\log x}}{\frac{x}{\log x}} = 0.$$

This does not mean that $\lim(\pi(x) - \frac{x}{\log x}) = 0$. Not at all. It just means that the discrepancy the size of the difference $\pi(x) - \frac{x}{\log x}$ is "negligible" when compared to $\frac{x}{\log x}$, as x grows to infinity. In other words, even if $|\pi(x) - \frac{x}{\log x}|$ would grow to infinity, its rate of growth is inferior to the rate of growth of $\frac{x}{\log x}$. What was said for $\frac{x}{\log x}$ may be repeated for the logarithmic integral. Again, the discrepancy $|\pi(x) - \text{Li}(x)|$ is negligible in respect to $\text{Li}(x)$, as x grows to infinity. Task 3 amounts to a comparison between $|\pi(x) - \frac{x}{\log x}|$ and $|\pi(x) - \text{Li}(x)|$, and to find out which function $\frac{x}{\log x}$ or $\text{Li}(x)$ approximates $\pi(x)$ more tightly.

Task 4 was out of reach for Gauss. It was beyond his means to come up with any other function to approximate $\pi(x)$.

Eric. I do not give a negative note to Gauss. It is normal that he could not prove what he had guessed. It was — if correct, which I don't know yet — a brilliant and penetrating discovery. Only a genius could formulate it.

P.P. The Prime Number Theorem is correct. We shall see another time the story of the P.N.T. and a few of its consequences. Gauss was far in advance over the technical resources that mathematicians had at that time. Much had to be invented in order to face the challenges of the proof. So it is no wonder that Gauss could not prove the theorem. When we meet next time, I'll have a difficult choice.

Notes About Karl Friedrich Gauss (1777–1855)

Born in Braunschweig (Germany), the young Gauss showed his remarkable gift for mathematics very early on. Recognition by the Duke of Braunschweig earned him — coming from an uneducated and poor family — the possibility of studying at a gymnasium and to attend university at Göttingen. Gauss excelled in classics, just as he did in mathematics and sciences. While still in his teens, Gauss made important discoveries in number theory. At university he got acquainted with the work of Fermat, Euler, Lagrange and Legendre, among others. In 1801, his *opera magna*, the book *"Disquisitiones Arithmeticae"* was published. It is considered one of the most influential mathematics books ever published. Gauss developed the theory of congruences, the law of quadratic reciprocity, the theory of binary quadratic forms, the theory of period of roots of unity and cyclotomy, including the possibility of construction by ruler and compass of regular convex polygons — for example with 17 sides. And the many more fundamental theories and ideas make this book, even today, a basic source for research. Gauss represented complex numbers by points in the plane (also done independently by Argand), he proved, as did d'Alembert, the fundamental theorem of algebra, stating that polynomials with coefficients which are complex numbers, have all their roots also complex numbers. In connection with prime numbers on empirical grounds — at age 15 — Gauss stated the fundamental Prime Number Theorem, which had to wait a century for its proof.

Today, Gauss is nicknamed the Prince of Mathematics, but he should receive high praise for his work in Astronomy and Physics. In mathematics, besides number theory, Gauss made seminal work in differential geometry leading eventually to non-Euclidean geometry and topology, on differential equations, on the study of complex numbers and the theory of functions of a complex variable and a variety of other topics. Among Gauss's important papers in astronomy, he invented a method which allowed one, with just three observations, to determine the orbit of a celestial body. In this way, the planetoid Ceres, discovered earlier and thereafter lost, could be again relocated. Other fields of research of Gauss included mechanics, crystallography, optics capillarity, and electromagnetism.

Notes About Johann Martin Zacharias Dase (1824–1861)

Dase was a prodigious mental calculator who remains unequalled to this day. Among his incredible feats, in 1844 Dase calculated 200 decimal digits of the number π. He also produced a 7-figure natural logarithm table for integers up to 1 005 000. Another important calculation was a table of factors of integers, in particular the determination of prime numbers, from 7 000 000 up to 10 000 000. Death prevented him from completing this task, which had been requested by Gauss.

22

Gathering Forces

P.P. Before the battle a general surveys his troops. Will they be strong enough to overcome the enemy?

Here we have to gather forces to attack the Prime Number Theorem. I need to explain many ideas which played a role in the proof of the Prime Number Theorem. It is a mixture of topics which at first appear unrelated to each other, but eventually come together and produce the support to the method which was successful in the proof of the Prime Number Theorem. So it will be a question of series, of the complex numbers — for us a new kind of numbers — and...

Paulo interrupted Papa Paulo with a restrained, but still visible exasperation:

Paulo. More numbers! I thought that with real numbers which measure any segment, and areas, we were through with numbers. And now you come with these complex numbers, which I guess, are unreal.

P.P. Yes, they were first called "imaginary". We shall need the analytic functions and many things that I will try to explain, as well as it is possible.

Series

P.P. There is no problem to add finitely many numbers. Sometimes we even have formulas. Remember, if $n \geq 1$ and $a \neq 1$, then $1 + a + a^2 + \cdots + a^{n-1} = \frac{a^n - 1}{a - 1} = \frac{1 - a^n}{1 - a}$.
But in other times we want to find the sum of infinitely many numbers.
Question: What happens when $a \geq 0$ and I want to evaluate the infinite sum $1 + a + a^2 + \cdots + a^{n-1} + a^n + \cdots$?

Paulo. If $a = 1$ we get infinity for the sum; of course the same is true if $a > 1$.

Eric. If $a = 0$ we get 1.

Paulo. Don't act smart. I rather listen to what you feel about the positive numbers a for which the sum $1 + a + a^2 + \cdots + a^n + \cdots$ is finite.

Eric. For me it is intuitive that if a is close to 0 then the sum $1 + a + a^2 + \cdots + a^n + \cdots$ cannot be too big.

Paulo. Let us take $a = \frac{1}{2}$ to see what happens. You begin with 1 and add $\frac{1}{2}$, you are now halfway before reaching 2. You add $\frac{1}{4}$, which is half of what was missing to reach 2 and you are halfway before 2, that is, you are halfway between $1 + \frac{1}{2}$ and 2. If you keep repeating the same, you are always less than 2, but getting closer and closer to 2. So I would say, without hesitation that $1 + \frac{1}{2} + \left(\frac{1}{2}\right)^2 + \left(\frac{1}{2}\right)^3 + \cdots$ is equal to 2.

P.P. Your reasoning was perfect. You can do the same for any number a such that $0 < a < 1$. The sum $1 + a + a^2 + \cdots = \frac{1}{1-a}$. And the proof is as follows:

$$1 + a + a^2 + \cdots + a^{n-1} + a^n + \cdots = \lim_{n \to \infty} \left(1 + a + a^2 + \cdots + a^{n-1}\right)$$

$$= \lim_{n \to \infty} \frac{1 - a^n}{1 - a} = \frac{1 - \lim_{n \to \infty} a^n}{1 - a}.$$

But $0 < a < 1$, so given $N > 0$, $\frac{1}{a} > 1$. As I have said once — remember? — when n is large enough $\frac{1}{a^n} = \left(\frac{1}{a}\right)^n > N$, hence $a^n < \frac{1}{N}$. This means exactly that $\lim_{n \to \infty} a^n = 0$.

Eric. This example is excellent. Now we know that sometimes an infinite sum has a finite value...

Papa Paulo interrupted, to say:

P.P. If the sum is finite we say that the sum is *convergent*.

Eric continued:

Eric. Other times the infinite sum has an infinite value...

New interruption:

P.P. The sum is now called *divergent*.

Paulo. Papa Paulo, I feel that you have illustrated well with these examples some important concepts. Can you spell out the definition? No detours.

P.P. OK, here it goes. Let $a_1, a_2, a_3, \ldots$ be an infinite sequence of numbers. This sequence of numbers defines the *series* $a_1 + a_2 + a_3 + \cdots + a_n + \cdots$. The *partial sums* of the series $a_1 + a_2 + a_3 + \cdots$ are the numbers $s_1 = a_1$, $s_2 = a_1 + a_2$, $s_3 = a_1 + a_2 + a_3, \ldots,$ $s_n = a_1 + a_2 + \cdots + a_n$ etc. The following cases may happen:

(1) the sequence of partial sums has a limit A which is a number (not infinity). In this case we say that the series $a_1 + a_2 + a_3 + \cdots$ is *convergent* and A is the sum of the series.
(2) the sequence of partial sums has limit infinity, in this case, the series is said to be *divergent*.
(3) the sequence of partial sums does not have a limit and again the series is said to be *divergent*.

An easy example for the third case is given by the uninteresting series where $a_1 = 1$, $a_2 = -1$, $a_3 = 1$, $a_4 = -1$, $a_5 = 1$, $a_6 = -1$ etc. Now $s_1 = 1$, $s_2 = 0$, $s_3 = 1$, $s_4 = 0$, $s_5 = 1$, $s_6 = 0, \ldots$.

Paulo. All clear in my sky, thanks. Let us see — I think it is interesting — what happens for the series $1 + \frac{1}{2} + \frac{1}{3} + \cdots + \frac{1}{n} + \cdots$. By the way, is there a short way to write the series?

P.P. A short way is $\sum_{n=1}^{\infty} \frac{1}{n}$. The sign Σ (sigma) means "sum"; sum of what? Of all the numbers $\frac{1}{n}$, where $n = 1, 2, 3, \ldots$, that is, exactly $1 + \frac{1}{2} + \frac{1}{3} + \cdots + \frac{1}{n} + \cdots$. Now the evaluation of the sum. Our heart is oscillating between convergence or divergence with an infinite sum.

Paulo. To decide I have to find the partial sums $s_1, s_2, s_3 \ldots$ to see how they are increasing and get a feeling if the limit of the partial sums is finite or infinite.

Eric. If you compare with $\sum_{k=0}^{\infty} \frac{1}{2^k} = 1 + \frac{1}{2} + \frac{1}{4} + \cdots$ every summand of this series is among the summands of $\sum_{n=1}^{\infty} \frac{1}{n}$, so this one may really grow to infinity. But the terms $\frac{1}{n}$ are smaller and smaller. So... I don't have a guess.

P.P. This is an instance when calculations could help to guess if the sum is finite or infinite. I guessed that you would ask me for these calculations. I came prepared (*I look at my notes*):

$$s_1 = 1, s_2 = s_1 + \frac{1}{2} = \frac{3}{2} < 2,$$

$$s_3 = s_2 + \frac{1}{3} = \frac{11}{6} < 2,$$

$$2 < s_4 = s_3 + \frac{1}{4} = \frac{25}{12} < 3.$$

I am not showing all the calculations. The partial sums grow very slowly. Here is an example: the smallest index n such that $s_n > 10$ is $n = 12367$.

Eric. The partial sums grow so slowly that I wonder if there will be one which is greater than 1000. Each time we are adding smaller and smaller natural numbers. At best, I am afraid from the looks, that it would take an incredible amount of computation to go beyond 1000.

P.P. A bit of clever thought does it. Group the summands

$$1 + \frac{1}{2} + \left[\frac{1}{3} + \frac{1}{4}\right] + \left[\frac{1}{5} + \frac{1}{6} + \frac{1}{7} + \frac{1}{8}\right] + \left[\frac{1}{9} + \frac{1}{10} + \cdots + \frac{1}{16}\right] + \cdots.$$

You see that the sum is greater than

$$1 + \frac{1}{2} + \frac{2}{4} + \frac{4}{8} + \frac{8}{16} + \cdots = 1 + \frac{1}{2} + \frac{1}{2} + \frac{1}{2} + \frac{1}{2} + \cdots.$$

We conclude that the sum $\sum_{n=1}^{\infty} \frac{1}{n} = \infty$.

After a short stop to impress that sometimes a clever thought wins over massive calculations, Papa Paulo continued:

P.P. Do you know what Euler did with the divergent series $\sum_{n=1}^{\infty} \frac{1}{n}$?

Paulo. If I were Euler I would throw it in a Saint Petersburg garbage. What could anyone do with a divergent series?

P.P. Euler used a divergent series to prove that there are infinitely many primes. Just for the pleasure of having a proof different from Euclid's. Euler proceeded as I show now:

Proof of Euclid's theorem by Euler: Assume that there are only finitely many primes, say denoted $p, q, \ldots, s$. We have:

$$\frac{1}{1 - \frac{1}{p}} = 1 + \frac{1}{p} + \frac{1}{p^2} + \frac{1}{p^3} + \cdots$$

$$\frac{1}{1 - \frac{1}{q}} = 1 + \frac{1}{q} + \frac{1}{q^2} + \frac{1}{q^3} + \cdots$$

$$\vdots$$

$$\frac{1}{1 - \frac{1}{s}} = 1 + \frac{1}{s} + \frac{1}{s^2} + \frac{1}{s^3} + \cdots$$

Multiplying the left-hand sides, one obtains the number

$$\frac{1}{1 - \frac{1}{p}} \times \frac{1}{1 - \frac{1}{q}} \times \cdots \times \frac{1}{1 - \frac{1}{s}}.$$

Multiplying the numbers on the right-hand side one gets a sum of numbers $\frac{1}{n}$. Each number n is a product of powers of primes $p, q, \ldots, s$ (all primes or just some of them). Since $p, q, \ldots, s$ are all the primes, any natural number n appears only once — this time Euler appealed to the other theorem of Euclid, namely that every positive integer is, in a unique way, a product of powers of primes. So on the right, one gets $\sum_{n=1}^{\infty} \frac{1}{n}$ (not written in the same order — but from the theory of series with positive terms, the order does not affect the sum). So look what Euler got: left-hand side finite, right-hand side $\sum_{n=1}^{\infty} \frac{1}{n} = \infty$. This is absurd, so there are infinitely many primes. q.e.d.

Eric, curious, said:

Eric. Papa Paulo, do you have more to tell about similar series?

P.P. This was the situation around that time. The question was if the series $\sum_{n=1}^{\infty} \frac{1}{n^2}$ is convergent and, if so, to evaluate the sum.

Eric. We saw that the series $\sum_{n=1}^{\infty} \frac{1}{n}$ is divergent, but also that the partial sums grow slowly, slowly, slowly. In the new series $\frac{1}{n^2}$ is far less than $\frac{1}{n}$ (when n is big) so most likely the partial sums of $\sum_{n=1}^{\infty} \frac{1}{n^2}$ would remain bounded. This is my guess.

P.P. Perfect guess. It is in fact rather easy to prove that the series $\sum_{n=1}^{\infty} \frac{1}{n^2}$ is convergent.

Eric. I think I know how to do it:

$$1 + \left(\frac{1}{2^2} + \frac{1}{3^2}\right) + \left(\frac{1}{4^2} + \frac{1}{5^2} + \frac{1}{6^2} + \frac{1}{7^2}\right) + \left(\frac{1}{8^2} + \cdots + \frac{1}{15^2}\right)$$
$$+ \cdots < 1 + \frac{2}{2^2} + \frac{4}{4^2} + \frac{8}{8^2} + \cdots = 1 + \frac{1}{2} + \frac{1}{4} + \frac{1}{8} + \cdots = 2.$$

So the partial sums of the series $\sum_{n=1}^{\infty} \frac{1}{n^2}$ are increasing, but always less than 2. Hence they have a limit which is the sum of the series.

P.P. The difficult problem that many people tried unsuccessfully to solve was to find the sum of the series. Euler showed that $\sum_{n=1}^{\infty} \frac{1}{n^2} = \frac{\pi^2}{6}$ (here $\pi = 3.14...$).

Euler wanted to find all real numbers $s > 1$ such that $\sum_{n=1}^{\infty} \frac{1}{n^s}$ is a convergent series. If s and s' are real numbers, if $1 < s < s'$ and s is good then so is s'. He was able to prove that for every $s > 1$ the series $\sum_{n=1}^{\infty} \frac{1}{n^s}$ is convergent. He did not try to compute explicitly the sum of the series, except when s is an even integer and $s \geq 2$ — in these cases he found very beautiful expressions. But he gave a name to the sum: $\zeta(s)$ (read zeta of s).

Paulo. Why "zeta"? I suppose by the looks that it is a Greek letter.

P.P. He had done many calculations and probably had used the other letters of the alphabet. But what is important, now he had the function $s \mapsto \zeta(s)$, defined for every $s > 1$. What I'll say now, I cannot prove for you (not having explained the appropriate notions in analysis), but it is very intuitive. If s is increased very little to s', then $\frac{1}{n^s}$ is also decreased a certain small amount to $\frac{1}{n^{s'}}$ and $\zeta(s)$ decreases also by a certain small amount to $\zeta(s')$. The function $\zeta(s)$ has no jumps, in other words, it is continuous.

Eric. From what you said, the domain of the function zeta is the set of real numbers $s > 1$ and the range of the function is the set of all real numbers $t > 0$. The function zeta is continuous, it is decreasing, $\lim_{s \to \infty} \zeta(s) = 1$ and $\lim_{s \to 1+0} \zeta(s) = \infty$ (read limit when s tends to 1 from the right). But what did Euler do with his zeta function?

P.P. He could prove
$$\sum_{n=1}^{\infty} \frac{1}{n^s} = \prod_{p} \frac{1}{1 - \frac{1}{p^s}}$$

(for $s > 1$). This formula contains the new symbol $\prod$, which indicates a product. Here there are infinitely many factors, one for each prime p. Just like for sums of series, there is a theory for infinite products which is very intuitive.

Paulo. It can only be as I'll say now. Write $p_1, p_2, p_3, \ldots, p_n, \ldots$ the sequence of primes in increasing order. Let

$$P_n = \prod_{k=1}^{n} \frac{1}{1 - \frac{1}{p_k^s}}.$$

Then

$$\prod_{p} \frac{1}{1 - \frac{1}{p^s}} = \lim_{n \to \infty} P_n.$$

P.P. That is it. The formula proved by Euler has on one side all primes, on the other side all positive integers. It reflects the fundamental theorem of Euclid on unique factorization.

P.P. We shall see one day how Riemann manipulated this idea, by using complex numbers. So it is time for the new numbers.

Complex Numbers

P.P. With real numbers you can measure segments. And you can find the square roots of any positive real number.

Paulo. I know the rule of signs of multiplication: positive times positive is positive, negative times negative is positive, while negative times positive is negative. So the square of a real number is not negative.

P.P. You said it right. Thus, -1 is not the square of any real number. Number systems are enlarged to make possible computations which were impossible in the smaller number system. An imaginary number i was invented with the express property that $i^2 = -1$. But one has also to consider all combinations $a + bi = a + ib$, where a and b are real numbers.

These are the *complex numbers*. The *real part* of $a+bi$ is the real number a, the *imaginary part* is bi. The *conjugate* of $a+bi$ is $a-bi$.

Paulo. This definition was so simple!

P.P. Now I give to you the definitions of the operations, using numerical examples, and also many properties of the complex numbers.
Sum: $(3+5i)+(-2+3i) = 1+8i$.
Subtraction: $(3+5i)-(-2+3i) = 5+2i$.
Multiplication: $(3+5i)\times(-2+3i) = 3\times(-2)+3\times 3i+5i\times(-2)+(5i)\times(3i) = -6+9i-10i+15i^2 = -21-i$.
Note: I did what anyone would do, and put $i^2 = -1$.

Division: $\dfrac{3+5i}{-2+3i} = \dfrac{(3+5i)\times(-2-3i)}{(-2+3i)\times(-2-3i)} = \dfrac{9-19i}{13} = \dfrac{9}{13}-\dfrac{19}{13}i$.

To perform the division, it was necessary to multiply numerator and denominator by the conjugate of the denominator. This fact was used: $(a+bi)(a-bi) = a^2+b^2$ is a positive real number (when $a+bi \neq 0$).

I continue saying things which are very simple. Every real number is a complex number, the zero complex number is just equal to the zero real number 0.

We define the *absolute value* of $z = x+yi$ as being $|z| = \sqrt{x^2+y^2}$. It is a positive real number, except when $z = 0 : |0| = 0$. Note that if $z = x$ is a real number then $|z|$ is the same as in the definition of absolute value of a real number. Other properties, easy to prove are: $|zz'| = |z|\times|z'|$ and $|z+z'| \leq |z|+|z'|$ for any complex numbers z and z'.

Eric. The real numbers could be represented by the points of the line, and all the points correspond to real numbers. Where can one represent imaginary numbers yi and arbitrary complex numbers $x+yi$?

Paulo answered:

Paulo. You must use a second line, the y-line for the imaginary numbers.

P.P. To represent $x + yi$ you do as follows. A horizontal line is the x-line, a vertical line is the y-line. These lines cross in a right angle at a point P_0 which represents the complex number 0. To represent $z = x + yi \neq 0$, draw the parallel to the y-line containing x, draw the parallel to the x-line containing yi. These two lines meet at a point P_z which represents $x + yi$. This point may also be called $x + yi$. The absolute value $|z| = \sqrt{x^2 + y^2}$ is the measure of the segment joining P_0 to the point P_z which represents $x + yi$. Taking $z \neq 0$ from $\left(\frac{x}{|z|}\right)^2 + \left(\frac{y}{|z|}\right)^2 = 1$ there exists a real number θ such that $\cos\theta = \frac{x}{|z|}$ and $\sin\theta = \frac{y}{|z|}$. In fact, θ is not unique, it can be replaced by $\theta + 2k\pi$, for any integer k. So we decide to make a choice for θ. We take $-\pi < \theta \leq \pi$ and call this chosen θ the *argument* of $z \neq 0$. The notation is $\theta = \arg(z)$.

Eric. Agreed. This is baby's trigonometry.

P.P. So we may write every non-zero complex number z in the form $z = |z|(\cos\theta + i\sin\theta)$, where $\theta = \arg(z)$. We note that $|\cos\theta + i\sin\theta| = 1$. This is the *polar form* of z. It is immediate to see that if $z = tu$, where t is real and positive and u is a complex number with $|u| = 1$ then $t = |z|$ and $u = \cos\theta + i\sin\theta$, where $\theta = \arg(z)$. Indeed $|z| = |tu| = |t| \times |u| = |t| = t$, so $|u| = \cos\theta + i\sin\theta$.

Paulo. You told us what the absolute value of the product zz' of two complex numbers z, z' is. Are you going to tell what is the argument of a non-zero product?

P.P. Yes. For this purpose it is convenient to introduce a "normalized sum" in the set of real numbers y such that $-\pi < y \leq \pi$. This is how I define it. Let $-\pi < y, y' \leq \pi$. If $-\pi < y + y' \leq \pi$ then $y + y'$ is the normalized sum of y and y'. If $y + y' > \pi$ then $y + y' - \pi$ is the normalized sum of y and y', because $0 < y + y' - \pi \leq \pi$. If $y + y' < -\pi$, then $y + y' + \pi$ is the normalized sum of y and y', because $-\pi < y + y' + \pi \leq 0$.

P.P. This is exactly what one has to do. And the usual properties of the sum are still valid.

Paulo. Do you use a special sign for the normalized sum? Like $n\underset{+}{+}$ or $\underset{+}{N}$?

P.P. One could, one should. But, it is better to realize from the context when the sum is normalized. Nobody will have any difficulty with this point. Before I tell you what is the argument of $zz' \neq 0$, I still need to refresh your memory. Remember these formula?
$$\cos(\theta + \theta') = \cos\theta \cos\theta' - \sin\theta \sin\theta'$$
$$\sin(\theta + \theta') = \sin\theta \cos\theta' + \cos\theta \sin\theta'.$$
You got these in school.

Papa Paulo proceeded:

P.P. Here is the expression for the argument of $zz' \neq 0$:
$\arg(zz') = \arg(z) + \arg(z')$ (the sum on the right-hand side is normalized).

Proof: $zz' = |z|(\cos\theta + i\sin\theta)|z'|(\cos\theta' + i\sin\theta')$ where $\theta = \arg(z)$ and $\theta' = \arg(z')$. Then $zz' = |z| \times |z'|((\cos\theta\cos\theta' - \sin\theta\sin\theta')) + i(\cos\theta\sin\theta' + \sin\theta\cos\theta') = |zz'|((\cos(\theta + \theta') + i\sin(\theta + \theta'))$. On the other hand, let $\theta'' = \arg(zz')$, so $zz' = |zz'|(\cos\theta'' + i\sin\theta'')$. So $\cos\theta'' = \cos(\theta+\theta')$ and $\sin\theta'' = \sin(\theta+\theta')$. Therefore, $\theta'' = \arg(zz')$ is the normalized sum of θ and θ'. q.e.d.

Functions, Limits, Series and All That Stuff

Eric. Papa Paulo, I suppose you are going to do the same as you did when the only numbers we knew were the real numbers. Are you going to introduce functions of complex numbers?

P.P. Yes.

Eric. Are you going to talk abort domain, range, graphs of the new functions?

P.P. Yes.

Eric. It must be very similar to what we saw for functions of real numbers. Why don't you just say, "We consider functions defined on complex numbers with values which are complex numbers: $z \mapsto f(z)$. The domain and the range are defined as for functions of real numbers."

P.P. This is exactly what I would have said.

Paulo. What about the graph?

P.P. The complex numbers $z = x + yi$ in the domain of the function $f(z)$ are represented as points of a plane. The complex numbers $f(z) = u + ti$ are also represented as points of another plane. To plot a graph we would need two planes, so four dimensions. We live in a 3-dimensional space. Graphs cannot be drawn. We have to use our imagination, our eyes are not useful.

Paulo. Then the study of complex functions is for blind mathematicians? (*Paulo knew that he was just being funny.*) But it is true that there have been blind mathematicians, like Euler was for 15 years at the end of his life.

Eric added:

Eric. There have been deaf musicians, like Beethoven.

P.P. And one-armed pianists, as Wittgenstein became after a war injury. Back to my functions, it is important to clearly define the notion of a limit. The functions which intervene in our discussion have domain D satisfying the following condition: if z_0 is in D then D contains a neighborhood of z_0, that is, there exists an integer $N > 0$ such that if $|z - z_0| < \frac{1}{N}$ then z is in D. We say that the complex number z_0 is arbitrarily close to D if for every integer $N > 0$ there exists z such that $|z - z_0| < \frac{1}{N}$.

Let z_0 be in the domain of $f(z)$ (or arbitrarily close to the domain of $f(z)$). To define the limit of $f(z)$ as z approaches z_0, it is not required that the domain of $f(Z)$ contains z_0, but it contains all complex numbers sufficiently close to z_0. The concept of limit is the same: the complex number w_0 is the limit of $f(z)$ as z approaches z_0 when $f(z)$ is arbitrarily close to w_0 when z is sufficiently close to z_0, $z \neq z_0$. This is expressed with inequalities as follows: First we assume that the positive integer N_0 is such that if $0 < |z - z_0| < \frac{1}{N_0}$, then z is in the domain of $f(z)$. Next, we impose that for every positive integer M there exists a positive integer $N > N_0$ such that if $z \neq z_0$ and $|z - z_0| < \frac{1}{N}$, then $|f(z) - w_0| < \frac{1}{M}$. We denote this fact by $\lim_{z \to z_0} f(z) = w_0$.

Paulo. For real functions you defined infinite limits. How about here?

P.P. Very simple: $\lim_{z \to z_0} f(z) = \infty$ when for every positive integer M there exists a positive integer $N > N_0$ such that if $z \neq z_0$ and $|z - z_0| < \frac{1}{N}$, then $|f(z)| > M$. If $\lim_{z \to z_0} f(z) = \infty$, the absolute value $|f(z)|$ may approach infinity at different rates depending on the function. If $\lim_{z \to z_0} (z - z_0) f(z)$ is neither 0 nor infinity, we say that $f(z)$ has a *pole of order* 1 at z_0. I don't need to say — you guessed — there are similar definitions for poles of order 2, 3, etc.

Paulo. Papa Paulo, you still did not say how to define $\lim_{z \to \infty} f(z) = w_0$ or ∞. But I can guess it myself.

P.P. Then do it on your own. It is more important to talk about analytic functions.

Eric. Surprise! Are you paying for my trip to Rio?

Paulo. Don't forget Paulo. I heard that the city is marvelous and I would also like to taste barbecued jararaca. After cooking, it is not poisonous.

P.P. Suppose you are already in Rio. Raise your heads. Towering at over 800 meters, on top of the mountain Corcovado, there is a huge statue of the Christ. Arms wide stretched. He is blessing the *cariocas*.

Assume that $\lim_{z \to z_0} \frac{f(z) - f(z)}{z - z_0}$ exists (and it is not infinity). This limit which depends only on $f(x)$ and on the z_0, is denoted by $f'(z_0)$ and it is called the *derivative* of $f(z)$ at z_0. If the derivative of $f(z)$ exists at every z_0 in the domain D of $f(z)$, we say that $f(z)$ is an *analytic function*. The theory of analytic functions is taught in the second year of a mathematics university curriculum. All functions in our discussion will be analytic.

Papa Paulo declared that he did not intend to describe systematically the properties of analytic functions. Nevertheless, in the following theorem, analytic functions would be needed.

P.P. Let $f(z)$ be an analytic function on the domain D, let $g(z)$ be an analytic function on the domain E. Assume that D and E have common points and that $f(z) = g(z)$ in all these common points. Then there is a unique analytic function $h(z)$, having domain H, which is the union of D and E, and such that if z is D, then $h(z) = f(z)$ and if $z \in E$, then $h(z) = g(z)$.

So, $h(z)$ is the analytic continuation of the functions $f(z)$ and $g(z)$.

Eric. What comes next?

P.P. I want to define the complex logarithmic function, the complex exponential function and the complex power functions.

Eric. You had already the functions with the same names defined as real numbers and now you want to define functions with values as complex numbers. Is there going to be any mixup?

P.P. Yes, at the beginning, until I clarify things. Let us agree that — just temporarily — I use the notation $R \log$ for the logarithmic function of real numbers which I had defined earlier. Remember? If $1 < x$, then $R \log x = \int_1^x \frac{dt}{t}$ and if $0 < x < 1$, then $R \log x = -R \log \frac{1}{x}$.

Now I define the logarithm of any complex number $z = x + yi \neq 0$:

$$\log z = R \log |z| + i \arg(z).$$

Thus, if $z = x$ is a non-zero real number, then $\log x = R \log |x|$. Hence if x is a positive real number, then $\log x = R \log x$. Therefore we can abandon the notation $R \log$.

We obtain, from the definition:

$$\log 1 = 0, \qquad \log(-1) = i\pi,$$
$$\log i = \frac{\pi}{2} i, \qquad \log(-i) = -i\frac{\pi}{2}.$$

The domain of the logarithmic function is the set of all non-zero complex numbers. The range of the logarithmic function is the set

of all complex numbers $z = x+yi$ where $-\pi < y \leq \pi$. On this set we consider the normalized sum: $(x+yi)+(x'+y'i) = (x+x')+(y+y')i$ where $x + x'$ is the usual sum of real numbers and $y + y'$ is the normalized sum. With this warning:
$$\log(zz') = \log z + \log z'.$$

Paulo. Papa Paulo, don't prove it. I will.

$$\begin{aligned}\log(zz') &= \log|zz'| + \arg(zz')i \\ &= \log|z| + \log|z'| + (\arg(z) + \arg(z'))i \\ &= \log|z| + \arg(z)i + \log|z'| + \arg(z')i = \log z + \log z'.\end{aligned}$$

q.e.d.

It was straightforward?

P.P. It was straightforward, which means that at every moment in the proof, there is only one thing to do, so the proof cannot go wrong — it goes forward in a straight way. And now the exponential function. Did I say it already? If $\log z = \log z'$ then $z = z'$?

Paulo. It comes from the unique polar form of any non-zero complex number.

P.P. Yes. Now, I shall define the exponential function. Let $w = t + ui$ be any complex number. There is a unique integer k such that $-\pi < u - 2k\pi \leq \pi$.

So $w - i2k\pi i$ is in the range of the logarithmic function. Hence there exists a unique non-zero complex number z such that $\log z = w - 2k\pi i$.

Then we define: $\exp w = z$.

It follows easily that if $z \neq 0$ then $\exp(\log z) = z$ and $\log(\exp w) = w - 2k\pi i$, where k was defined above. The domain of the exponential function is the set of all complex numbers, while the range of the exponential function is the set of all non-zero complex numbers.

I use, just for one minute, the notation $R \exp$ for the exponential function of real numbers. If $w = t$ is a real number, then $\exp t = R \exp t$.

Paulo. I hate the R's and I want to abandon this notation. Can you give me the proof?

Eric. This time I do it. Let $w = t$ be a real number. Then $R \exp t$ is a positive real number, so $\log(\exp t) = t = R \log(R \exp t) = \log(R \exp t)$. From this, (I used already this fact) $\exp t = R \exp t$.

P.P. Paulo and Eric, abandon the superfluous R-notation. Now I indicate another property of the exponential function
$$\exp(w + w') = (\exp w)(\exp w).$$
As a special case, if $z = x + yi$ then $\exp z = (\exp x)(\exp yi)$.

Paulo. I don't care to prove it. Straightforward in my opinion. But I am curious about $\exp(yi)$.

P.P. There is a very nice expression in terms of trigonometric functions:
$$\exp(yi) = \cos y + i \sin y.$$

Proof: Let k be the unique integer such that $-\pi < y - 2k\pi \leq \pi$. We have $|\cos y + i \sin y| = 1$ and $\arg(\cos y + i \sin y) = y - 2k\pi$. Then $\log(\cos y + i \sin y) = \log 1 + \arg(\cos y + i \sin y)i = (y - 2k\pi)i$. Hence $\exp(yi) = \cos y + i \sin y$. q.e.d.

A few seconds later, Eric said:

Eric. Papa Paulo, you still have to define the power function.

P.P. It all starts with powers of positive integers. For example, if $a = 3$, then $3^0 = 1$ and if n is a positive integer, we know that is 3^n : 3 multiplied with itself n times. We had defined 3^{-n} to be $\frac{1}{3^n}$. We also defined 3^r for any rational number r. Next we introduced the power 3^x for any real number x. In the above definitions, 3 may be replaced by any positive real number y. From $\log(y^x) = x \log y$, it follows that $y^x = \exp(x \log y) = e^{x \log y}$.

Paulo. I think you are saying all this because you are convinced that we have forgotten what you said earlier.

P.P. I know that you did not forget. My reason is to make acceptable the definition of the complex power function. If w is a non-zero

complex number, then the power w^z is defined as follows:
$$w^z = \exp(z \log w).$$
If w is a given complex number, $w \neq 0$, the function $z \mapsto w^z$ deserves the name of *power function*, w is the *base* and z is the *exponent*.

If z is a given complex number, the function $w \mapsto w^z$ (defined for all $w \neq 0$) is not the same as the power function.

Eric. Are you going to tell the properties of these functions?

P.P. Well, why not. But don't ask me to prove them.

Paulo. I know, you will say "It is straightforward." You are sort of funny, Papa Paulo. When the proof is hard, you skip it, and I understand the reason — we would not be able to follow the proof. But when the proof is straightforward you also skip it, saying "It is too boring." Yet, you gave several proofs, some very easy. What is your guideline?

P.P. We are having discussions. I am not teaching you a university level course. My aim is always to explain the ideas and make the developments plausible. The proofs which were shown are mainly for the purpose of illustration.

Eric. Can you list the main properties of powers?

P.P. They are what you would expect:
$$w^{z+z'} = w^z\, w^{z'}$$
$$w^{zz'} = (w^z)^{z'}$$
$$(ww')^z = w^z\, w'^z.$$

At this point, it became clear that we had gathered forces.

23

─∘◦✦◦∘─

The After "Math" of Gauss

As Eric and Paulo joined him in the afternoon, Papa Paulo began to speak:

P.P. Gauss could not prove his Prime Number Theorem, but the challenge to prove the theorem mobilized many mathematicians for almost the whole 19th century. The proof was considered to be so difficult that the one who would succeed would no doubt become immortal. This was not meant literally, but just that the name would not be forgotten in the history of mathematics.

Paulo. Could anyone do the proof out of the blue, or was it the result of gradual progress?

P.P. The "blue" does not prove theorems. The ideas and technical advances which we saw in our last meeting were the fabric of the proof. But I want first to tell you a significant theorem in that direction. You will see that it resembles the Prime Number Theorem, but there are essential differences, which I will explain.

Eric. Done by whom?

P.P. Remember Eratosthenes, a name which you could not spell? How about this one? PAFNUTY LVOVICH CHEBYSHEV?

Paulo. Too hard for me. I'll call him Paffy. What did he prove?

P.P. Paulo, I don't know if "Paffy" is your idea of an affectionate name. I hope it is not to show lack of respect. He was eminent and proved in 1850 his beautiful theorem:

Theorem: *For every integer $N > 0$ there exists an integer $N_1 > 0$ (which depends on N) such that if $x > N_1$, then*

$$\left(C - \frac{1}{N}\right)\frac{x}{\log x} < \pi(x) < \left(C' + \frac{1}{N}\right)\frac{x}{\log x}$$

where

$$C = \frac{1}{30}\log\frac{2^{15} \times 3^{10} \times 5^6}{30} = 0.92120\ldots$$

and $C' = \frac{6}{5}C = 1.10555\ldots$.

What the theorem of Paffy — oops — Chebyshev says is that $\pi(x)/(\frac{x}{\log x})$ is between $C - \frac{1}{N}$, and $C + \frac{1}{N}$ as soon as x is sufficiently large. What the theorem does not say is how the values of $\pi(x)/(\frac{x}{\log x})$ behave. Jumping like crazy or tending to a limit? A second feature of the theorem is that if it can be proved that there is a limit of $\pi(x)(\frac{x}{\log x})$, then $\lim \pi(x)/(\frac{\pi(x)}{\log x}) = 1$, as x tends to infinity.

Eric. Papa Paulo, you explained Chebyshev's theorem very well, and we can see what is missing to reach the Prime Number Theorem. Did the proof of Chebyshev's theorem involve techniques which were unknown to Gauss?

P.P. The proof did not appeal to methods which were unknown to Gauss. It was a clever proof. If Gauss had wanted to prove this theorem, I am convinced that he could have done it.

Paulo. Are there already interesting consequences of Paffy's theorem?

P.P. Indeed, there is one which I love. A "voyeur" of tables of primes named Bertrand, noted that between any number N and $2N$ there is at least one prime.

Paulo. Quite incredible. Give me two minutes. I want to look at my table.

After two minutes, Paulo said:

Paulo. I confirm what Bertrand observed. I have even noticed that for many integers N there is a prime much closer to N than $2N$;

also there may be many primes between N and $2N$. I wonder how the proof goes.

P.P. Chebyshev got Bertrand's empirical result as a consequence of his theorem. It is clear that if $N = 2$, 3 or 4, there is a prime between N and $2N$. If $N \geq 5$, the inequalities of Chebyshev imply that $\pi(2N) - \pi(N) > \frac{1}{3}\frac{N}{\log N} > 1$, so there is a prime between N and $2N$.

Paulo. Very neat. If anyone, such as that mentally prodigious Dase was preparing a table of primes, having reached the prime p he would be sure that the next one was before $2p$.

Eric. Useless... if p was about 4 million, to say that the next prime is before 8 million is without any interest. What he would need is to know that the next prime would be much closer to p.

P.P. Actually, the Bertrand observation, proved by Chebyshev, is a result towards the understanding of gaps between primes.

Eric. Will you say anything more about this? I know that sometimes the gap is just 2, like between 3 and 5, 5 and 7, 11 and 13, etc. By Bertrand's observation, the gap after the prime p is less than p.

P.P. Keep this in mind. But now I come back to my main aim...

Paulo. P.N.T.

P.P. Not long after Chebyshev, Riemann published his famous paper on primes. It was full of ideas and opened a royal way leading eventually to the proof of the P.N.T.

Paulo. With beautiful flowers bordering the sides of this boulevard?

P.P. Metaphorically, yes, but also with some steep inclines. Do you know what I meant with the word "metaphorically"? Well, I say it right now. The way was traced to advance in the research of the distribution of primes. The flowers are meant to be the beautiful theorems and formulas which Riemann had to discover and use.

Eric. Enough of flowers. Facts!

P.P. Remember that Euler defined the zeta function $\zeta(s) = \sum_{n=1}^{\infty} \frac{1}{n^s}$, for real numbers $s > 1$. He proved the product formula which involves the primes and reflects Euclid's fundamental theorem of factorization:

$$\sum_{n=1}^{\infty} \frac{1}{n^s} = \prod_{p} \frac{1}{1 - \frac{1}{p^s}}, \quad \text{for} \quad s > 1.$$

This product is now called the *Euler product* for the zeta function.

Riemann thought: "I shall consider the series $\sum_{n=1}^{\infty} \frac{1}{n^s}$ when s is a complex number, not only a real number $s > 1$. I will find the values of s for which the series is convergent and define, like Euler did, the zeta function $\zeta(s)$." This new function (he thought with obvious pride) came to be named the *Riemann zeta function*.

Eric. Was Riemann successful?

P.P. Yes. $\sum_{n=1}^{\infty} \frac{1}{n^s}$ is convergent for every complex number s with real part $\text{Re}(s) > 1$. And it defines the zeta function $\zeta(s)$ on that region of the complex plane. Just like for the Euler zeta function, the Riemann zeta function is exempt of misbehavior — it is an analytic function. Riemann continued his thoughts: "Now I have my zeta function with a much larger domain than the one of Euler's, which was only defined for the real numbers $s > 1$. Euler had his product formula. I can also get my product formula. My function is therefore related to primes, I have more room to maneuver. It has a bigger domain, it is an analytic function, so it will be possible to apply more powerful theorems from the rich theory of analytic functions."

Paulo. Adventurous Riemann.

P.P. Next, Riemann said to himself, "I will extend the most I can, the domain of my zeta function, so I will have even more room to maneuver."

Eric. But you said that the zeta series $\sum_{n=1}^{\infty} \frac{1}{n^s}$ converges only when $\text{Re}(s) > 1$ and we already know that it is divergent when $0 < \text{Re}(s) \le 1$. So what do you mean by "extend the domain"?

P.P. Riemann showed that there exists an analytic function which has domain equal to the whole complex plane, except the point $s = 1$, and which is equal to $\sum_{n=1}^{\infty} \frac{1}{n^s}$, when $\operatorname{Re}(s) > 1$. Actually this was not overly difficult. Of course, he still used the notation $\zeta(s)$ for the extended function — which is the unique possible analytic function providing the required extension.

Papa Paulo continued describing the achievements of Riemann.

P.P. I don't know if the next step was a surprise discovery or a brilliant guess of Riemann. This is what he did: he multiplied $\zeta(s)$ with a negative power of π and a value of a classical function studied by Euler, the gamma function $\Gamma(s)$. He found a remarkable and precious symmetry for the values of his product

$$\frac{\Gamma\left(\frac{s}{2}\right)}{\pi^{\frac{s}{2}}} \cdot \zeta(s) = \frac{\Gamma\left(\frac{1-s}{2}\right)}{\pi^{\frac{1-s}{2}}} \cdot \zeta(1-s).$$

This functional relation tells that what happens at the left of the vertical line containing the real number $s = \frac{1}{2}$ has its counterpart at the right of that line.

It is also possible to evaluate $\zeta(0) = -\frac{1}{2}$. The gamma function has infinite values at $s = -1, -2, -3, \ldots$. It follows that the zeros of $\zeta(s)$ are of two kinds:

First kind: the trivial zeros at $s = -2, -4, -6, \ldots$.

Second kind: the zeros which are in the critical strip — this is the set of all complex numbers s such that $0 \leq \operatorname{Re}(s) \leq 1$.

Paulo. I suppose that it would be too difficult to define and give the main properties of the gamma function. But can you at least give a proof of what you said about the zeros of the zeta function?

P.P. This I can do. I will need the fact that $\Gamma(s)$ is never equal to 0. I also need that $\sum_{n=1}^{\infty} \frac{1}{n^s} \neq 0$ when $\operatorname{Re}(s) = 1$. This follows from the product formula, because each factor $\frac{1}{1-\frac{1}{p^s}} \neq 0$. Next, if $\operatorname{Re}(s) < 0$ then $\operatorname{Re}(1-s) > 1$, so the right-hand side in the functional equation is not equal to 0. But $\Gamma(s)$ has value infinitely at $s = -1, -2, -3, \ldots$. This forces that $\zeta(-2) = 0$, $\zeta(-4) = 0$, $\zeta(-6) = 0, \ldots$.

Paulo. If you have all you need, the proof is easy.

Eric. Let me try to recapitulate what was done, beginning with Euler.

(a) Euler defined his zeta function $\zeta(s)$ for real numbers $s > 1$, and he proved the product formula.

(b) Riemann defined $\zeta(s) = \sum_{n=1}^{\infty} \frac{1}{n^s}$ for $\operatorname{Re}(s) > 1$ and showed that it is an analytic function.

(c) Riemann extended the zeta function to the whole complex plane as an analytic function, except at $s = 1$, because $\zeta(1) = \infty$.

(d) Riemann found the functional equation for the function $\frac{\Gamma\left(\frac{s}{2}\right)}{\pi^{\frac{s}{2}}} \zeta(s)$; it does not change when s is replaced by $1 - s$.

(e) Riemann determined the trivial zeros of the zeta function and deduced that all the other zeros are in the critical strip.

What else did Riemann do?

P.P. What happens as s tends to 1 was not hard to settle, namely $\lim_{s \to 1}(s - 1)\zeta(s) = 1$. It is the same as saying that $\zeta(s)$ and $\frac{1}{s-1}$ have the same growth rate as s tends to 1. The point $s = 1$ is therefore a pole of order 1 of $\zeta(s)$.

It was Riemann's inner feeling that the location of the zeros of $\zeta(s)$ in the critical strip had much to do with the distribution of primes. I don't know what prompted him, but he did it. He spelled out what is today called the *Riemann Hypothesis*. All the zeros of the Riemann zeta function which are in the critical strip are on the critical line, namely the vertical containing $s = \frac{1}{2}$. Riemann felt with good reason that if his hypothesis was assumed to be true, it would have many consequences about the distribution of primes.

Eric. Papa Paulo, you made us believe — we do! — that Riemann had tremendous new ideas. I suppose that mathematicians of his time and later wanted to use Riemann's methods to prove the Prime Number Theorem.

P.P. This is exactly what happened. But the challenge was so great that it became (I told you already) folklore that he who would prove the P.N.T. would become immortal.

Eric. Tell us what happened.

P.P. Not one, but two mathematicians, in the same year 1896, each one unaware of the other, succeeded to prove the P.N.T. Many facts from the theory of analytic functions were needed. It was established that a certain region inside the critical strip did not have any zero of the Riemann zeta function.

Eric. And that was, if I understood, in agreement with the feeling of Riemann.

P.P. To be precise, there exists $c > 0$ and $t_0 > e^{2c}$ such that $\zeta(s)$ has no zero in the region of all complex numbers $s = u + it$ satisfying the inequalities

$$\begin{cases} 1 - \dfrac{c}{\log t_0} \leq u \leq 1 & \text{when} \quad 0 \leq t < t_0 \\ 1 - \dfrac{c}{\log t} \leq u \leq 1 & \text{when} \quad t_0 \leq t. \end{cases}$$

Pick some values for c and t_0 and draw this region to see its shape.

This was a crucial point in the proof of the Prime Number Theorem.

Paulo. You said that two mathematicians proved the theorem, thus becoming immortal. Are their names now forgotten?

P.P. They will never be forgotten. And you know, they lived so long — past the age of 95 — that it was expected that they would indeed be immortal. As for their names...

Papa Paulo seemed to have forgotten, so he said:

P.P. Let me look at my notes to get their names. Oh yes, here they are. One was Hadamard (a Frenchman), the other one was de la Vallée Poussin (a Belgian).

Eric. This was a mathematical feat. You did not indicate how the proofs went. I believe that it was very hard. But you know what amazes me?

Eric continued:

Eric. Hard facts from analytic functions were apparently unavailable to prove something about prime numbers, which are integers after all. A question, Papa Paulo. Do you think that this is right? In your opinion, would it be possible to have a proof without analytic functions and all that hard stuff?

P.P. During the period after Riemann, up to the first half of the 20th century, mathematicians were unanimous in thinking that analytic functions could not be dispensed to obtain the proof. Proofs, simpler than the two original ones were published, but always appealing to analytic methods. Until...

Eric. I guess someone managed to liberate the proof from analytic functions.

P.P. Not someone, but again two mathematicians, Erdős and Selberg. They independently and in direct competition published in 1949 long sought different proofs of the Prime Number Theorem, devoid of any recourse to the theory of analytic functions.

The story of the P.N.T. is an intellectual saga, a major advancement in the knowledge of prime numbers. I suppose that you are now burning with curiosity to see the consequences of the Prime Number Theorem and other important facts about the distribution of primes.

Suddenly, Papa Paulo glanced at his watch and exclaimed:

P.P. Hey! Look at the time. It is the moment to stop.

Paulo added:

Paulo. If we at least had a good dinner...

And with no inhibition, he continued:

Paulo. I will settle for soup, an Italian main course (no pizza) and for my sweet tooth, you know my taste for pies.

His charming smile could not be resisted, so Papa Paulo said:

P.P. You are lucky. Nana prepared soup for two, to last for five meals. But you each eat like four: $2 \times 5 = 10 = 2 \times 1 + 2 \times 4$. You see my calculation. We can finish all the soup today. One bowl for Nana, one for Papa Paulo and four for each monster.

Eric. Which kind of soup?

P.P. Squash, carrots and green lime.

Eric. It is a strange combination.

Paulo. Nana made, it is good. The soups of Nana are great.

P.P. The main dish will be homemade tortellini, filled with an authentic Italian creamy cheese, of ricotta type.

Eric. My favorite stuff, and I tell you, for dessert my number 1 is ice cream, a pie is my number 2.

P.P. There will be 12 or 21, whatever order you like to eat them. I am teaching you, fellows, a new kind of arithmetic: $12 = 21$.

Paulo. Which kinds are 1 and 2?

P.P. Nana told me: Apricot pie, French double vanilla ice cream (all natural, no chemicals added).

Paulo. Any Champagne? I like brut, very cold.

Papa Paulo did not answer, but he said:

P.P. Usually before a meal, many people thank the Higher Powers for the bread they are eating. Today we are not having bread. It is Italian food with a touch of French. Philosophers come to mind; not the ones thriving on nothingness or writing about the Absolute. By far I prefer the chef-philosophers, with whom I find a brain and stomach resonance. Jean-Pierre Coffe gave a profound thought: "You are a better person if you eat good food."

Papa Paulo had not finished stating the aphorism and measuring its impact, when everyone heard the joyful voice of Nana "Good food is at table, good soup is steaming."

We all thought: Let us become better persons. At the table three pairs of eyes, bright and happy were saluting the mastery of Nana.

Notes About Pafnuty Lvovich Chebyshev (1821–1894)

Chebyshev, born in Okatovo, Russia, moved early in life to Moscow. He received tutoring in mathematics from P. N. Pogorelski, a famous mathematics teacher and author of numerous successful texts for high school levels. Chebyshev was well prepared when he entered Moscow University in 1837, at age 16. Very much influenced by Brashman, he acquired a solid knowledge of analysis and mechanics.

His entire research body would show the ambivalence and fruitful interplay between pure and applied mathematics. He was the author of important papers in probability theory and also began the study of general orthogonal polynomials. It is also agreed that Chebyshev is the founder of the Russian school of approximation theory. In applied mathematics, Chebyshev's interests turned towards the limit of $\pi(N)/(n/\log n)$.

Nevertheless, he was not able to show the existence of the limit, thus falling short of the proof of the Prime Number Theorem conjectured by Gauss and eventually proved by de la Vallée Poussin and Hadamard.

Most of Chebyshev's career was spent as a professor at the St. Petersburg University. He taught about the new steam and other machines, but he was also the inventor of numerous engineering devices including a specially designed bicycle for women.

In number theory, Chebyshev published in 1852 a paper containing the proof of Bertrand's conjecture, namely that there is always a prime between an integer, greater than 1, and its double. This was a consequence of the inequalities for $\pi(N)$ (the number of primes less or equal to N). In recognition of his achievements the mathematical community bestowed on Chebyshev many honors. He became a member of the most prestigious academies and received numerous honorary degrees.

Notes About Georg Friedrich Bernhard Riemann (1826–1866)

Born in Breselenz, Germany, Riemann died at age 40 of tuberculosis. But in his short life he produced many important papers!

In 1846 he entered the University of Göttingen, where he heard lectures by Stern and Gauss. One year later he went to the University of Berlin, and there enjoyed closed contacts with Dirichlet and Eisenstein.

His doctoral thesis (1851, under the supervision of Gauss), habilitation thesis and inaugural lecture (1854) were masterpieces. They concerned the theory of functions of a complex variable, what we now call *Riemann surfaces*. Already by then, Riemann had introduced important new ideas and proved the condition of integrability of functions, the representation by means of trigonometric series. In the lecture, he presented his results which concerned Riemannian geometry. This was the type of geometry which Einstein needed to develop his theory of relativity. Further work was devoted to the theory of Abelian integrals and topology.

Upon Riemann's election to the Berlin Academy in 1859, he presented his papers on the function $\pi(N)$, counting the number of primes less or equal to N. His main tool was the zeta function, which he considered as a function of a complex variable. Riemann proved it's analytic continuation to the whole plane, except the pole at the point 1. Then he established a functional equation, stated that the zeta function has infinitely many zeros with a real part between 0 and 1. He stated that probably, the real part of these non-trivial zeros is equal to $\frac{1}{2}$. This assertion constitutes the so-called Riemann Hypothesis. Its proof (or disproof) is one of the most difficult problems in mathematics today. Riemann was a giant in mathematics.

Notes About Charles Jean Gustave Nicolas de la Vallée Poussin (1866–1962)

De la Vallée Poussin, born in Louvain (Belgium), did not quite reach his 100th birthday. Yet, having proved the Prime Number Theorem — at the same time as Hadamard — in 1896, he was supposedly immortal. Well, if not his body, his name is. The proof of the Prime Number Theorem was the most important open problem since the statement had been conjectured by Gauss. The work of Riemann brought attention to the Riemann zeta function, but much remained to be done. Powerful ideas and fine analysis were needed by de la Vallée Poussin to prove the theorem. At that time, de la Vallée Poussin was just 30 years old, but he did not remain idle. The important book *"Cours d'Analyse"*, which had many editions, constantly updated with the inclusion of novel theorems, was a most influential work for the study of mathematical analysis. De la Vallée Poussin also wrote several other books treating topics in the forefront of research, like Lebesgue integration, approximation of real valued functions, logarithmic potential, etc.

Needless to say, de la Vallée Poussin was much honored, with elections to academies, honorary doctoral degrees, medals and invitations to address meetings.

Notes About Jacques Salomon Hadamard (1865–1963)

Born in Versailles, France, Hadamard did not quite reach the age of 100 (like de la Vallée Poussin). Having proved the Prime Number Theorem, he too should become immortal; his work is. The fact is that during his extensive career, Hadamard made so many important contributions in so many disciplines of mathematics, that there is general agreement that Hadamard should be considered one of the most important mathematicians of his time.

First in the entrance examination for the Ecole Normale Supérieure in 1884, Hadamard followed courses by such illustrious mathematicians as Tannery, Hermite, Goursat, Darboux, Appell. His doctoral thesis in 1892 was about singularities of functions of a complex variable. In the same year, he received the Grand Prix for his work on the distribution of primes which, among others, aimed at completing work as suggested by Riemann.

Hadamard proved the Prime Number Theorem in 1896 — the same year as de la Vallée Poussin — namely that $\pi(N)$ (the number of primes less or equal to N) grows as $\frac{N}{\log N}$ when N grows to infinity. This was the greatest achievement in analytic number theory and required extensive knowledge of the theory of analytic functions of a complex variable. Important as this result is, it may be said that it is only one jewel in his crown, considering the totality of outstanding and fundamental theorems that Hadamard was able to prove.

These range from estimates of sizes of determinants, the special class of the so-called Hadamard matrices, the well-posed problems as he called them, the foundations of functional analysis, but also geodesics of surfaces, differential equations, hydrodynamics, geometric optics, elasticity and even more. Hadamard also wrote a widely adopted textbook in two volumes about elementary geometry. His book on the psychology of mathematical invention was an original contribution to the understanding of the mind of mathematicians. Over 300 articles contain the fruits of Hadamard's research. Honors received were numerous and he deserved them without question. During the Dreyfus question, around the turn of the century, Hadamard took a leading and courageous position in the successful campaign for the rehabilitation of Dreyfus, whose condemnation was supported by forged documents of the accusation and anti-Semitic feeling.

Tragedy also hit Hadamard, when he lost two of his sons in World War I, another son in World War II, and a grandson in a mountain accident.

Hadamard's scientific achievements, work ethic, and integrity made him a towering figure and an example for succeeding generations.

Notes About Atle Selberg (1917–2007)

Selberg was born in Langesand, Norway. He studied in the University of Oslo where he was influenced by the work of the Norwegian school of number theory, especially Axel Thue and Viggo Brun. He studied Ramanujan with great interest and was deeply struck by a 1936 lecture by Hecke. In 1947 he moved to the United States and in 1950 had joined the Institute for Advanced Study in Princeton as a permanent member.

In 1949, Selberg proved the Prime Number Theorem using no analytic function theory, but just elementary methods. In the following year, he also gave an elementary proof for the corresponding theorem for primes in arithmetic progressions.

At the 1950 Cambridge (Massachusetts) International Congress of Mathematicians, Selberg received the Fields Medal. This was an acknowledgment not only of the elementary proofs just quoted, but of Selberg's invention of the large sieve, which is a generalization of Brun's sieve. This new technique enabled Selberg to make a substantial progress towards the proof of the Riemann Hypothesis. He showed that the set of non-trivial zeros of Riemann's zeta function which are on the critical line with abscissa $\frac{1}{2}$, has positive density.

Selberg's work has been always of the highest level. He has contributed many outstanding results in deep-lying topics in number theory, like automorphic functions, the trace formula for unimodular 2×2 real matrices and the Selberg zeta function. All this is work of central importance.

Notes About Pál Erdős (1913–1996)

Born in Budapest, Hungary, Erdős was very protected by his mother — no wonder; just weeks before his birth, two sisters had died in an epidemic of scarlet fever. World War I, the captivity of his father by Russian troops, the communist regime and right wing dictatorship, coupled with anti-Jewish laws (forerunner of Hitler's) was the atmosphere of Erdős' youth. In 1934, Erdős received his doctoral degree from the University of Budapest, and soon left for the United Kingdom. This was to be the beginning of his unceasing travel, which took him not only to Hungary (when it finally became possible) but also to Israel, The Netherlands, the United States, Canada, France and many other countries. Despite numerous highly advantageous offers, Erdős held only temporary positions. It was his working style to solve — by himself or with collaborators — difficult problems in the most elegant manner.

At age 18, he gave a transparent proof of Bertrand's conjecture (proved earlier by Chebyshev) that between N and $2N$ there is always a prime. In 1949, to the surprise of mathematicians and at the same time as Selberg, Erdős gave an elementary proof of the Prime Number Theorem using an entirely new method. It is estimated that Erdős is the author of over 1500 papers, in the areas of number theory, graph theory and combinatorics. Erdős was not a builder of theories, but an unequaled problem solver.

24

Primes After Dinner: Bad Dreams?

Dinner over, with time to relax and then we were back to primes. Nana was careful to stay far enough away to not even hear the word "prime". She had other major interests in people, not numbers.
Eric began:

Eric. In our discussion, it was a question of one of the tasks that Gauss wanted to complete. How about finding the best fitting continuous computable function to approximate $\pi(x)$?

P.P. Gauss had already the functions $\frac{x}{\log x}$ and $\mathrm{Li}(x)$. Riemann knew these functions, but he thought that they could not fit tightly. So Riemann defined another function. His idea was to make an appropriate combination of the logarithmic integral, computed at various real values. Do you remember the Möbius function?

Paulo. I vaguely remember that you once said the name "Möbius".

P.P. OK. The Möbius function is the following:
$\mu(1) = 1$
$$\mu(n) = \begin{cases} 1 \text{ if } |n| \text{ is the product of an even number of distinct primes} \\ -1 \text{ if } |n| \text{ is the product of an odd number of distinct primes} \\ 0 \text{ if } n = 0 \text{ or if } n \neq 0 \text{ and it is divisible by the square of a prime.} \end{cases}$$

Now I make a very simple remark. If x is a real number, $x \geq 1$, for every n, sufficiently large $2^n > x$, so $x^{1/n} < 2$ and the logarithmic integral of $x^{1/n}$, for all sufficiently large n is equal to zero.

With this remark, it is guaranteed that for every $x \geq 1$, the sum
$$\sum_{n \geq 1} \frac{\mu(n)}{n} \operatorname{Li}(x^{1/n})$$
has only finitely many summands which are not equal to 0.

Eric. 100% agreed. And then what?

P.P. Riemann defined what is now called the *Riemann function*:
$$R(x) = \sum_{n \geq 1} \frac{\mu(n)}{n} \operatorname{Li}(x^{1/n})$$
for all real numbers $x \geq 1$.

Eric. Now there are three functions competing for the title of "best fitting to $\pi(x)$". How far have computations been done?

P.P. There are not only computational results, but also theorems. For $\pi(x)$ and $\frac{x}{\log x}$ it has been proved that if $x \geq 11$ then $\pi(x) > \frac{x}{\log x}$. Computations show that $\pi(x) - \frac{x}{\log x}$ may be quite big. For example:
$$\pi(10^{21}) - \frac{10^{21}}{21 \log 10} = 446579871578168707.$$

Eric. And how big is $\pi(10^{21})$?

P.P. It is equal to: $\pi(10^{21}) = 21127269486018731928$. I am copying these numbers from my notes. So $\frac{\pi(10^{21}) - \frac{10^{21}}{21 \log 10}}{\pi(10^{21})}$ is around $\frac{1}{7}$. For bigger x this quotient shall become smaller, the limit being 0, even though the difference $\pi(x) - \frac{x}{\log x}$ grows bigger.

Paulo. I don't think that $\frac{x}{\log x}$ provides a good approximation for $\pi(x)$. How about $\operatorname{Li}(x)$ or $R(x)$?

P.P. In my notes I have $\operatorname{Li}(10^{21}) - \pi(10^{21}) = 597394254$ and $R(10^{21}) - \pi(10^{21}) = -86432204$. So for $x = 10^{21}$ it is the Riemann function which provides the best approximation to $\pi(x)$.

Eric. I have no clue on how to calculate the logarithmic integral and Riemann's function.

P.P. These questions are dealt with in computational number theory. Usually the functions are replaced by a series which provide a good approximation. For large values of x there are technical problems. I am not a specialist in these questions. It was laborious to obtain the exact values of $\mathrm{Li}(10^{21})$.

Eric. And how do we calculate $\pi(x)$ for very large values of x? It is clear that it is possible to calculate the value of $\pi(x)$, but you have already insisted that it should be done in polynomial time.

P.P. There is an old formula due to Legendre, obtained from an analysis of the sieve of Eratosthenes. For $N \geq 2$:

$$\pi(N) = \pi(\sqrt{N}) - 1 + \sum_d \mu(d) \left[\frac{N}{d}\right].$$

In this formula $\mu(d)$ denotes the value of the Möbius function, $\left[\frac{N}{d}\right]$ is the unique integer such that $\left[\frac{N}{d}\right] \leq \frac{N}{d} < \left[\frac{N}{d}\right] + 1$; finally in the $\sum_d$ there is one summand for each d = product of distinct primes p, where $p \leq \left[\sqrt{N}\right]$. The number of summands is equal to $2^{\pi[\sqrt{N}]}$, which is the number of sets of distinct primes $p \leq \left[\sqrt{N}\right]$. Too many summands for any practical use as soon as N is not small.

Paulo. Let me see. We know that $\pi(100) = 25$. So $\pi(10\,000) = \pi(100) - 1 +$ sum of 2^{25} summands. Horrible. Then how were you able to quote the value $\pi(10^{21})$?

P.P. A formula for $\pi(x)$ was invented by Meissel. It needs the knowledge of the primes $p \leq x^{\frac{1}{2}}$, the values of $\pi(y)$ for $y \leq x^{\frac{2}{3}}$, it also has a sum, but only of $\pi(x^{\frac{1}{2}}) - \pi(x^{\frac{1}{3}})$ summands. I don't say that it is easy to use such a formula, but it is manageable.

Paulo. I am starting to admire the computational people.

Eric. Now I like to ask: are there consequences of the Prime Number Theorem?

P.P. Of course, there are many, but I'll be brief, since I do not want to be involving you in questions which are too technical and definitely not of primordial importance (even though of prime importance).

Paulo. Give us one to begin.

P.P. I write $p_n \sim n \log n$ which means that $\lim_{n \to \infty} \frac{p_n}{n \log n} = 1$.

Paulo. Can I say something? The sequence of primes p_n grows to infinity faster than the sequence of natural numbers does, but not so fast as the sequence of squares grows. And it grows exactly like the sequence of numbers $n \log n$. Can I try to prove it?

P.P. Of course. It will be simple.

Paulo. You said that it is a consequence of the Prime Number Theorem. I have to calculate $\lim_{n \to \infty} \frac{p_n}{n \log n}$ and show that it is equal to 1. This is what I do:

We know that $\pi(p_n) = n$, so by the Prime Number Theorem

$$\lim_{n \to \infty} \frac{n \log p_n}{p_n} = \lim_{n \to \infty} \frac{n}{\frac{p_n}{\log p_n}} = 1.$$

Hence

$$\lim_{n \to \infty} \log \frac{n \log p_n}{p_n} = \log \lim_{n \to \infty} \frac{n \log p_n}{p_n} = 0.$$

So

$$\lim_{n \to \infty} (\log n + \log \log p_n - \log p_n) = 0.$$

Dividing by $\log p_n$ we have $\lim_{n \to \infty} \left(\frac{\log n}{\log p_n} + \frac{\log \log p_n}{\log p_n} - 1 \right) = 0$.

But we all know that $\lim_{n \to \infty} \frac{\log x}{x} = 0$; since $\lim_{n \to \infty} \log p_n = \infty$, then

$$\lim_{n \to \infty} \frac{\log \log p_n}{\log p_n} = 0.$$

Therefore $\lim_{n \to \infty} \frac{\log n}{\log p_n} + \lim_{n \to \infty} \frac{\log \log p_n}{\log p_n} - 1 = 0$, so $\lim_{n \to \infty} \frac{\log n}{\log p_n} = 1$.

In conclusion, $\lim_{n\to\infty} \dfrac{n \log n}{p_n} = \lim_{n\to\infty} \dfrac{n \log p_n}{p_n} \times \dfrac{\log n}{\log p_n} = 1.$

I could prove it!   q.e.d., q.e.d., q.e.d.

P.P. Bravo Paulo. Here is another consequence. Let N be a very large natural number, pick a number n such that $1 \le n \le N$. What is the probability that n is a prime?

This is very easy to solve. The probability is the quotient of the number of possible primes, which is $\pi(N)$, divided by N. For N large, by the Prime Number Theorem, this probability is

$$\dfrac{\pi(N)}{N} \sim \dfrac{\frac{N}{\log N}}{N} = \dfrac{1}{\log N}.$$

Paulo. Let me try one. If I pick a number n, $1 \le n \le 10^9$ the probability that n is a prime is $\dfrac{1}{\log 10^9} = \dfrac{1}{20.7} < \dfrac{1}{20}$.

Eric. Any more consequences of the P.N.T. which you find easy to understand?

P.P. Here is one. For every $n \ge 1$ let $d_n = p_{n+1} - p_n$, so d_n is the difference between the consecutive primes p_{n+1} and p_n. Then $\lim_{n\to\infty} \dfrac{d_n}{p_n} = 0$.

Eric. It is my turn. This is very easy to see:

$$\dfrac{d_n}{p_n} = \dfrac{p_{n+1} - p_n}{p_n} = \dfrac{p_{n+1}}{p_n} - 1.$$

But $\dfrac{p_{n+1}}{(n+1)\log(n+1)}$ has limit 1 and also $\dfrac{p_n}{n \log n}$ has limit 1. So

$$\lim_{n\to\infty} \dfrac{p_{n+1}}{p_n} = \lim_{n\to\infty} \dfrac{\dfrac{p_{n+1}}{(n+1)\log(n+1)} \times (n+1)\log(n+1)}{\dfrac{p_n}{n \log n} \times n \log n}$$

$$= \lim_{n\to\infty} \dfrac{n\left(1 + \frac{1}{n}\right) \log(n+1)}{n \log n}$$

$$= \lim_{n\to\infty} \dfrac{\log\left(n\left(1 + \frac{1}{n}\right)\right)}{\log n}$$

$$= 1 + \lim_{n\to\infty} \dfrac{\log\left(1 + \frac{1}{n}\right)}{\log n} = 1. \qquad \text{q.e.d}$$

P.P. Now we are touching on an aspect of the distribution of primes which we have not yet debated. If you have a prime p_n in your hands, how much bigger is p_{n+1}?

Paulo. Please review what we already know.

P.P. Sometimes $d_n = 2$, like when we have 3 and 5, 5 and 7, 11 and 13, 17 and 19 and in many more instances. We have seen that $d_n < p_n$, because $p_{n+1} < 2p_n$. Remember, this was stated by Bertrand and proved by Chebyshev. And we also proved, as a consequence of the Prime Number Theorem, that $\frac{d_n}{p_n}$ has limit 0.

Papa Paulo continued:

P.P. Would you expect the next fact? Give me any number N, as big as you want. Look at the N numbers $(N+1)! + 2$, $(N+1)! + 3$, $(N+1)! + 4, \ldots (N+1) + (N+1)$.

They are all composite because they are divisible respectively by $2, 3, 4, \ldots, N+1$. Let p_n be the largest prime smaller than $(N+1)! + 2$, so $p_{n+1} > (N+1)! + (N+1)$. Hence $d_n = p_{n+1} - p_n \geq N$. We have shown the difference of consecutive primes may be that arbitrarily large.

Eric. I am not surprised after your scare campaign about the differences d_n. Now, go ahead and tell us more about the erratic behavior of the sequence of d_n's.

P.P. It is true that there are infinitely many n such that $d_n > d_{n-1}$ and also infinitely many n such that $d_n < d_{n-1}$.

Specialists on the theory of prime numbers have made substantial efforts to understand the sequence of the d_n's. Their results are too technical, even to mention to you, and many unsolved problems are challenging the mathematicians. I will debate some of the mysteries another day.

It is midnight and all the primes we "absorbed" after dinner may give you bad dream.

25

Primes in Arithmetic Progression

After resting from the marathon discussion session, Papa Paulo had to choose the topic of the day. Just like being in a restaurant. The menu has many very appetizing dishes, from which you must choose one. You think: I had heavy food in the past days, so I'll settle for something lighter. Papa Paulo was in a similar mood in the face of primes. He wanted to reserve discussing the mysteries just before vacation.

P.P. It appears very suitable that today we talk about primes in arithmetic progressions.

Paulo. Tell us, what is an arithmetic progression.

P.P. It is better if I first give some examples:

- 1 3 5 7 9 11... — you begin with 1 and each time you add 2;
- 3 7 11 15 19... — you begin with 3 and each time you add 4.

You can begin with any integer a and each time you add d which you assume to be a positive integer. You get the arithmetic progression with *initial term a* and *difference d*. You can do many things with arithmetic progressions. The term in position n is $a + (n-1)d$. The sum of the first n terms in $a + (a+d) + (a+2d) + \cdots + (a+(n-1)d) = na + [1 + 2 + \cdots + (n-1)]d$.

The formula $1 + 2 + \cdots + (n-1) = \frac{(n-1)n}{2}$ is the one little Gauss found, when his teacher was mad and asked the kids to add $1+2+3+\cdots+100$. Do you remember? So $a + (a+d) + (a+2d) + \cdots + (a+(n-1)d) = na + \frac{(n-1)n}{2} \times d$. This is simple and nice but not the topic of our discussion. We shall focus on this question. Are there always prime numbers in any arithmetic progression? If so, how many?

Eric. If your arithmetic progression is 1 2 3 4 5 6..., which is allowed, then there are infinitely many — in fact all — primes in the arithmetic progression.

Paulo continued in a mocking manner — you could see in his eyes:

Paulo. If the arithmetic progression is 1 3 5 7 9... then all the primes except 2 are in the arithmetic progression.

Eric added a useful remark:

Eric. If there exists a number $q > 1$ which divides both a and d, then q divides each term of the arithmetic progression — which therefore has no primes with the possible exception of a when $q = a$ is itself a prime. Remember: this appeared in the sieve of Eratosthenes.

P.P. So it remains to examine the situation when a and d are coprime. It is good to begin with simple arithmetic progressions. Let $d = 4$ and $a = 1$ 5 9 13 17... these numbers are congruent to 1 modulo 4. I show:

Theorem: *There exist infinitely many primes p such that $p \equiv 1 \pmod{4}$.*

Proof: We assume the contrary and we shall reach a contradiction. Let $p_1, p_2, \ldots, p_r$ be the only existing primes which are congruent to 1 modulo 4. Let $N = (2p_1 p_2 \ldots p_r)^2 + 1$ and let p be a prime number which divides N. So p is neither equal to 2, nor to $p_1, \ldots p_r$. We show that $p \equiv 1 \pmod 4$. Indeed $(2p_1 \ldots p_r)^2 + 1 \equiv 0 \pmod p$ so $-1 \equiv (2p_1 p_2 \ldots p_r)^2 \pmod p$. You see that -1 is a square modulo p. Remember! This means that $p \equiv 1 \pmod 4$. This is trouble, because p is now still another prime in the arithmetic progression. Absurd, and the proof is completed. q.e.d.

Paulo. It reminds me of Euclid, but using the result about -1 being a square modulo p.

Eric. And how about the arithmetic progression 3 7 11 15 19 ... .

P.P. The same result holds.

> **Theorem.** *There exist infinitely many primes $p \equiv 3 \pmod 4$.*

> **Proof:** The proof is again by reduction to an absurd. We assume that $p_1, p_2, \ldots p_r$ are the only primes which are congruent to 3 modulo 4. Let $N = 2p_1 p_2 \ldots p_r + 1$. Whether r is even or odd $2p_1 p_2 \ldots p_r \equiv 2 \pmod 4$ so $N \equiv 3 \pmod 4$. Therefore there exists a prime p which divides N and such that $p \equiv 3 \pmod 4$. You know why? Because a product of primes congruent to 1 modulo 4 is congruent to 1 modulo 4. But $p \neq p_1, p_2, \ldots, p_r$, so p is a new prime congruent to 3 modulo 4. Absurd. q.e.d.

P.P. I want to go back to the primes $p \equiv 1 \pmod 4$. They have a very interesting property discovered by Fermat.

Paulo. Really? Or was it no more than a guess?

P.P. There are guesses and there are guesses. But this time there is no doubt that Fermat had a good idea for the proof of his theorem. Missing steps were completed by Euler.

Paulo. Are you telling us the proof? Or letting us find it?

Papa Paulo hesitated and finally said.

P.P. You are not Fermat. I'll tell the proof; it is very pretty.
   The theorem of Fermat:

> **Theorem.** *Let p be a prime. If $p \equiv 1 \pmod 4$, then p is the sum of two non-zero squares. If $p \equiv 3 \pmod 4$, then p is not the sum of two non-zero squares.*

> **Proof:** First we assume that m and n are positive integers such that $m^2 + n^2$ is a prime p. So m and n are non-zero. If m and n are even then $m^2 + n^2$ is divisible by 4, so it is not a prime. If m and n are odd, say $m = 2m_0 + 1$, $n = 2n_0 + 1$, then $m^2 + n^2 = (2m_0 + 1)^2 + (2n_0 + 1)^2 = (4m_0^2 + 4m_0 + 1) + (4n_0^2 + 4n_0 + 1)$,

and $m^2 + n^2 =$ (multiple of 4) $+2$ is even, so it is not a prime. It remains the case where $m = 2m_0$, $n = 2n_0 + 1$ (or vice-versa), so $p = m^2 + n^2 = 4m_0^2 + (4n_0^2 + 4n_0 + 1) =$ (multiple of 4) $+1$.

In conclusion, if the prime $p \equiv 3 \pmod 4$, then p is not the sum of two non-zero squares.

Now we assume that $p \equiv 1 \pmod 4$. We use the fact that -1 is a square modulo p, so there exists a non-zero integer x such that $1 \leq x \leq p - 1$ and $-1 \equiv x^2 \pmod p$. Hence there exists m, $1 \leq m \leq p - 1$ such that $mp = x^2 + 1$.

Thus there exists an integer m, such that $1 \leq m \leq p - 1$ and mp is the sum of two squares. Of all these integers m, there is one which is the smallest possible. We call it m_0, so $1 \leq m_0 < p - 1$. If we show that $m_0 = 1$, then $p = m_0 p$ is the sum of two squares. What happens if m_0 is not equal to 1? Let $m_0 p = x^2 + y^2$. We divide x by m_0: $x = am_0 + r$ with $0 \leq r < m_0$. If $r \leq \frac{m_0}{2}$ we are happy and do nothing. If $\frac{m_0}{2} < r$ we write $x = (a+1)m_0 + r - m_0$ and now $-\frac{m_0}{2} < r - m_0$.

So, whatever be the case, we may write $x = am_0 + r$ where $-\frac{m_0}{2} < r \leq \frac{m_0}{2}$.

In the same way $y = bm_0 + s$ where $-\frac{m_0}{2} < s \leq \frac{m_0}{2}$.

It is forbidden to have $r = s = 0$. Indeed, if $r = s = 0$, then $m_0 p = x^2 + y^2 = m_0^2(a^2 + b^2)$, so $p = m_0(a^2 + b^2)$. $m_0 \neq 1$, then $m_0 = p$, but this is absurd. It follows that $0 < r^2 + s^2 \leq \frac{m_0^2}{4} + \frac{m_0^2}{4} = \frac{m_0^2}{2} < m_0^2$. Remember, m_0 divides $x^2 + y^2$ and

$$x \equiv r \pmod{m_0}, \qquad y \equiv s \pmod{m_0}$$

so $x^2 + y^2 \equiv r^2 + s^2 \pmod{m_0}$, therefore m_0 divides $r^2 + s^2$. Hence $r^2 + s^2 = m'm_0$ and $m' < m_0$ because $r^2 + s^2 < m_0^2$. We multiply $(m_0 p)(m'm_0) = (x^2 + y^2)(r^2 + s^2)$. This product, where the factors are the sum of two squares, is again the sum of two squares (no magic, just a little observation)

$$m_0^2 m' p = (x^2 + y^2)(r^2 + s^2) = (rx + sy)^2 + (ry - sx)^2.$$

If you don't believe, begin from the right:

$$(rx + sy)^2 = r^2 x^2 + s^2 y^2 + 2rsxy,$$
$$(ry - sx)^2 = r^2 y^2 + s^2 x^2 - 2rsxy,$$

so the sum is $r^2x^2 + s^2y^2 + r^2y^2 + s^2x^2 = (r^2+y^2)(r^2+s^2)$.

Where are we now? Oh, yes. We write

$$rx + sy = (x - am_0)x + (y - bm_0)y$$
$$= x^2 + y^2 - m_0(ax + by)$$
$$= m_0p - m_0(ax + dy).$$

Thus, m_0 divides $rx + sy$ and we write $rx + sy = m_0t$. We do a similar calculation:

$$ry - sx = (x - am_0)y - (y - bm_0)x = -m_0(ay - bx),$$

hence there exists an integer u such that $ry - sx = m_0u$. It follows that $m_0^2m'p = m_0^2(t^2+u^2)$, hence $m'p = t^2+u^2$ where $1 \leq m' < m_0$. But m_0 was the smallest possible with the property which m' also has. Absurd. q.e.d.

After hearing the proof, Paulo said:

Paulo. The proof was clever, not straightforward. I'd like to see examples, all primes $p \equiv 1 \pmod 4$, up to 100, written as a sum of two non-zero squares.

P.P. Easy:

$$5 = 2^2 + 1^2$$
$$13 = 3^2 + 2^2$$
$$17 = 4^2 + 1^2$$
$$29 = 5^2 + 2^2$$
$$37 = 6^2 + 1^2$$
$$41 = 5^2 + 4^2$$
$$53 = 7^2 + 2^2$$
$$61 = 6^2 + 5^2$$
$$73 = 8^2 + 3^2$$
$$89 = 8^2 + 5^2$$
$$97 = 9^2 + 4^2$$

Paulo. The numbers were small so it was easy. How difficult would it be for big numbers?

Eric. I also wonder: In how many ways can a prime $p \equiv 1 \pmod{4}$ be written as the sum of two squares?

P.P. These questions may be answered by studying the *Gaussian integers* $m + ni$, where m and n are any integers and i is the imaginary unit. Remember? $i^2 = -1$. The Gaussian integers are complex numbers. We can — and did — define the operations of sum, subtraction and multiplication of Gaussian integers. So it is possible to study the arithmetic of Gaussian integers. In particular, one can say when Gaussian integer divides another Gaussian integer.

Eric. You also have prime Gaussian integers?

P.P. Yes.

Eric. Is there also a theorem of unique factorization as a product of prime Gaussian integers?

P.P. Yes, this theorem is also true. But it needs a slight modification, which is natural for Gaussian integers.

Eric. Don't explain too much, but say a little bit about the modification.

P.P. I will discuss one numerical example. You will be able to see the new phenomena. The number 5 is a prime number, among the integers. But $5 = 5 + 0i$ is also a Gaussian integer and $5 = (2+i)(2-i)$, so 5 is not a prime Gaussian integer. It may be shown that $2+i$ and $2-i$ are prime Gaussian integers. The Gaussian integers $1, -1, i$ and $-i$ are divisors of 1, because $1 \times 1 = 1$ $(-1) \times (-1) = 1$, $i \times (-i) = 1$. They are called the *units*. A prime Gaussian integer multiplied with a unit is still a prime Gaussian integer. So $(-1) \times (2+i) = -2-i$, $i(2+i) = -1+2i$, $(-i)(2+i) = 1-2i$ are prime Gaussian integers; so are $-2+i, -1-2i$ and $1+2i$. And we also have the following decomposition of 5 as product of primes:

$$5 = (-2-i)(-2+i) = (1-2i)(1+2i) = (-1+2i)(-1-2i).$$

But these decompositions are considered to be "the same up to units" as $5 = (2+i)(2-i)$.

Paulo. I note, with some apprehension, that to solve some problems about integers one has to create new numbers like the Gaussian integers. Study their arithmetic, ask problems about the Gaussian integers. This creates still newer kinds of numbers. A Pandora's box.

P.P. This is very true. The numbers created belong to the so-called *algebraic numbers*. Their study is unavoidable for anyone wanting to deal with integers at a higher level of sophistication.

After a short pause to consider his words, Papa Paulo said:

P.P. Remember that author Miobnebir? He has written a fat book about algebraic numbers; hundreds of pages.

Papa Paulo stopped. He had mentioned Miobnebir's book saying that it was fat, but would not admit that it was well written, even excellent. It is clear that there is a problem between Papa Paulo and the other guy — jealousy?
Eric moved back to mathematics:

Eric. Papa Paulo, you proved that if $d = 4$ and $a = 1$ or $a = 3$, the corresponding arithmetic progressions contain infinitely many primes. The proofs were at an understandable level. Can you do something similar to other arithmetic progressions?

P.P. Yes, but I'll not show you. Take $d = 3$ and $a = 1$ or 2, take $d = 6$ and $a = 1$ or 5; take $d = 8$ and $a = 1, 3, 5$ or 7; take $d = 12$ and $a = 1, 5, 7$ or 11; take $d = 24$ and $d = 1, 5, 7, 11, 13, 17, 19$ or 23.
For all these there exist specific proofs which are not too hard.

Eric. So it works for certain small values of d.

P.P. There are also proofs for arbitrary $d > 2$ when $a = 1$ or $a = d-1$. I call your attention to this.

If d is a very large integer it is not obvious that there will even be a prime number of the form $1 + kd$ or $d - 1 + kd$. But the fact is that the following theorem is true:

If $d > 2$ there exists infinitely many primes $p \equiv 1 \pmod{d}$ and infinitely many primes $p \equiv -1 \pmod{d}$.

Paulo. Will you prove these facts?

P.P. I will prove that there exist infinitely many primes $p \equiv 1 \pmod{q^s}$, when q is a prime and $s \geq 1$.

Proof: Let $r \geq 0$ and assume that $p_1, p_2, \ldots, p_r$ are the only primes congruent to 1 modulo $d = q^s$. We shall show that there exists a prime $p \equiv 1 \pmod{q^s}$ and such that $p \neq p_1, p_2, \ldots, p_r$. This is a contradiction thus proving the theorem. Let $N = qp_1 \ldots p_r$, let $M = N^{q^{s-1}}$. We apply Newton's binomial formula and a trick:

$M^q - 1 = [(M-1) + 1]^q - 1 = (M-1)^q + \binom{q}{1}(M-1)^{q-1} + \binom{q}{2}(M-1)^{q-2} + \cdots + \binom{q}{q-1}(M-1)$. So $M - 1$ divides $M^q - 1$.

Since $M - 1 < M^q - 1$ there exists a prime p such that p divides $\frac{M^q - 1}{M - 1} = \frac{N^{q^s} - 1}{N^{q^{s-1}} - 1}$.

If p divides $M - 1$, then p divides $\binom{q}{q-1} = q$ so $p = q$, but q divides M, hence p divides 1, which is absurd. Thus p does not divide $M - 1 = N^{q^{s-1}} - 1$, but p divides $M^q = 1 = N^{q^s}$. So $N^{p^s} \equiv 1 \pmod{p}$ but $N^{q^{s-1}} \not\equiv 1 \pmod{p}$. Since p and N are coprime then by Fermat's theorem $N^{p-1} \equiv 1 \pmod{p}$. Let $p \geq 1$ be the smallest positive integer such that $N^{-t} \equiv 1 \pmod{p}$. Then t divides q^s but t does not divide q^{s-1}, so $t = q^s$. But $q^s = 1$ divides also $p - 1$, then $p \equiv 1 \pmod{q^s}$ with $p \neq p_1, \ldots, p_r$. This concludes the proof.

q.e.d.

Paulo. A very nice proof indeed, but valid only when d is a prime power and $a = 1$.

P.P. If d is not a prime power there is also a proof which is not too hard, but requires more background, so I shall not show it.

Eric. Are you forgetting the arithmetic progressions with arbitrary d and $a = d - 1$?

P.P. There is also an elementary proof, but this time it uses arithmetic properties of binary recurrences. It is not possible to show the proof in our discussion.

Eric. There is something funny, Papa Paulo. You said "elementary proofs" when d is arbitrary and $a = 1$ or $d - 1$. But you did not want to show these proofs. I conclude that they are not so easy and should not be called "elementary".

P.P. In contrast, the general case when d is arbitrary and $1 \leq a < d$, with $\gcd(a, d) = 1$ is incomparably more difficult. It was proved by Dirichlet in 1837. The proof required analytic methods; series which are similar to the zeta series, but using certain functions called the *characters modulo d*. Anyway, these are things that I am not discussing.

Eric. Will you at least state the theorem of Dirichlet?

P.P. For any real number $x > 1$, I denote by $\pi_{d,a}(x)$ the number of primes p such that $p \leq x$ and $p \equiv a \pmod{d}$. Recall that $\varphi(d)$ (the Euler function) is the number of integers a such that $1 \leq a < d$ and $\gcd(a, d) = 1$.

Paulo. From what you said earlier, $\varphi(d)$ is the number of arithmetic progressions with difference d which may contain one prime.

P.P. Exactly. Dirichlet proved: If $1 \leq a < d$ and $\gcd(a, d) = 1$, there are infinitely many primes $p \equiv a \pmod{d}$.

Later, with the method used to prove the Prime Number Theorem, it was shown that $\pi_{d,a}(x) \sim \frac{1}{\varphi(d)} \times \frac{x}{\log x}$. This is the same as saying that $\lim\limits_{x \to \infty} \frac{\pi_{d,a}(x)}{\frac{x}{\log x}} = \frac{1}{\varphi(d)}$.

Eric. Papa Paulo, aren't you making a mistake? The left-hand side has "a" present so it depends on the chosen arithmetic progression $a, a + d, a + 2d, \ldots$. But the right-hand side does not have "a". Isn't this strange?

P.P. You are sharp-eyed, but what I wrote is correct. It is true that the primes up to x are equally distributed in the $\varphi(d)$ arithmetic progressions. Not exactly for each x, but when x tends to infinity.

You can also say that if $1 \leq a, a' < d$ with $\gcd(a, d) = 1$ and $\gcd(a', d) = 1$, then $\lim\limits_{x \to \infty} \dfrac{\varphi_{d,a}(x)}{\varphi_{d,a'}(x)} = 1$.

You can also write $\lim\limits_{x \to \infty} \dfrac{\pi_{d,a}(x)}{\pi(x)} = \dfrac{1}{\varphi(d)}$ and say, in a suggestive way, that the density of primes in each arithmetic progression with difference d is equal to $\dfrac{1}{\varphi(d)}$.

Papa Paulo was visibly happy with the presentation of the important theorem of Dirichlet on primes in arithmetic progressions. He was particularly elated to have been able to hide all the technical innovation needed to prove the theorem in its general form. Perhaps he should have said that the ideas of Dirichlet to prove the theorem have found application — sometimes in modified form — in a large number of problems in the theory of prime numbers. But he didn't say it. Instead, he continued:

P.P. My next comments have a double aim. Primo, to give more facts about primes in arithmetic progressions. Secondo, to call your attention about what can be proved, versus what can be illustrated by numerical examples.

Eric. You told us many times: Look at your table of primes, find patterns and then say "Conjecture such and such should be true." Put it on the Internet and expect that someone will say that your conjecture is false, or that a proof will be posted, or even that there is no reaction from the thousands of addicted people.

P.P. And the statement under scrutiny will be known as "Eric's conjecture". But there is another method of research. You may ask yourself questions which look reasonable. You translate these questions into statements. For example: "For every positive integer k there is a set of k numbers which satisfy a required property." Someone may prove the statement using sophisticated methods of analytic number theory. But the proof does not allow us to determine explicitly the sets with k numbers having the required property. There is a proof but no examples (when k is not too small). In other cases the statement depends on a positive real number x. It is possible to prove that there exists $x_0 > 0$ such that the statement is true when $x > x_0$. According to the specific problem, x_0 may be effectively computable, but usually big. Or x_0 is proved to exist but it is unspecified. Again, there is a proof but no numerical example.

Eric. I don't like your mysterious description of this alternative research method. I like what is concrete. Tell us *something*.

P.P. OK. This will be convincing. Here there are three legitimate questions:

Question 1: Is it true that for every $k > 2$ there exists an arithmetic progression (or even infinitely many arithmetic progressions) of k terms which are prime numbers?

Question 2: Is it true that for every $k > 2$ there exists an arithmetic progression (or even infinitely many arithmetic progressions) of k terms which are consecutive primes?

Question 3: Is it true that for every odd prime p there exists $k > 1$ such that $p, p+k, p+2k, \ldots, p+(p-1)k$ are primes?

It is clear that these questions are related. A positive answer to Question 2 implies a positive answer to Question 1. Similarly, a positive answer to Question 3 guarantees that there is one arithmetic progression which answers Question 1.

Paulo. I would not ask any of the three questions. But some people are more inquisitive than I am. Papa Paulo, what are you going to say?

P.P. Question 1 is of the utmost difficulty. Just recently the following theorem was proved:

Theorem. *There exist arbitrarily long arithmetic progressions whose terms are prime numbers.*

The proof does not allow us to find an example concretely. Will anyone ever find one million primes in arithmetic progression? Go to sleep; this will never happen. Now we know that if the answer to Question 2 is positive, the proof has to be hard. But you will be surprised with what I say now. In a future discussion of mysteries, I will explain that if a certain conjecture is assumed to be true, then the answer to Question 2 is positive.

Eric. Quite amazing!

Paulo. Unbelievable! I am eager to know why.

P.P. Hold your horses and you'll see it. Anyway, 3, 5, 7 are three consecutive primes in arithmetic progression. 251, 257, 263, 269 are four consecutive primes in arithmetic progression.

Eric. Papa Paulo, give me twenty — no more than twenty — consecutive primes in arithmetic progression.

P.P. The longest known string of consecutive primes in arithmetic progressions contains ten terms. The initial prime is

$$p = 100996972469714247637786655587969840329509324689190041803603417758904341703348882159067229719,$$

and the difference of the arithmetic progression is 210.

This problem is still unsolved for Question 3, but the answer is positive for small primes. Here is what is known:

$p = 3$:   3   5   7
$p = 5$:   5   11   17   23   29
$p = 7$:   7   157   307   457   607   751   907
$p = 11$: $11 + h \times 1536160080821194590$ (for $h = 0, 1, \ldots, 10$)
$p = 13$: $13 + h \times 918821194590$ (for $h = 0, 1, 2, \ldots, 12$)
$p = 17$: $17 + h \times 341976204789992332560$ (for $h = 0, 1, 2, \ldots, 16$).

No numerical examples are known for $p > 17$.

It is believed that the answer to Question 3 will be positive. From the large amount of calculations already required for small primes like $p = 11, 13, 17$, even if a proof is found, it will be hard to exhibit numerical examples.

Eric. I always thought that what I explained to you just now was the right way to make discoveries. Now you come with another method. And worse: the proofs — if you can find them — may be very hard and you cannot even see numerical examples. What kind of number theory is this? I hate analytic number theory.

P.P. Don't hate what you can do with great effort. Learn to respect the inherent difficulty of problems about primes.

Paulo did not want to extend this controversy. He asked:

Paulo. Papa Paulo, did you not say — I think you did: If d is big it is not clear how big the smallest prime $p \equiv 1 \pmod{d}$ has to be. So I ask: let $1 \leq a < d$, with $\gcd(a,d) = 1$. How soon can you find a prime $p \equiv a \pmod{d}$?

P.P. Your question, based on my earlier passing comment, touches an important aspect. For a careful discussion, I introduce the following notation. If $1 \leq a < d$ with $\gcd(a,d) = 1$, let $p(d,a)$ be the smallest prime $p \equiv a \pmod{d}$. Let $p(d)$ be the maximum of all the numbers $p(d,a)$. We know that this number exists and we wish to show that it cannot be bigger than some quantity which depends on d; so the same is true for each prime $p(d,a)$. Instead of saying what we know right away, I will state what could be conceivable estimates for $p(d)$. Then I will discuss each one. False? True? Uncertain?

Eric. This is a good plan and will help us to understand and to know what to expect. From your precautions, Papa Paulo, I suspect that the question is thorny.

P.P. Thorny, but beautiful like a rose.

Estimate 1: There exists an integer $M \geq 1$ such that $p(d) < d + M$ for every $d > 1$.

Estimate 2: There exists an integer $M \geq 1$ such that $p(d) < Md$ for every $d > 1$.

Estimate 3: There exists an integer $M \geq 1$ such that $p(d) < M d (\log d)^2$ for every $d > 1$.

Estimate 4: $p(d) < d^2$ for every sufficiently large d.

Estimate 5: $p(d) < d^{5.5}$ for every sufficiently large d.

Paulo. Papa Paulo, you made a list of conceivable estimates for $p(d)$. It is clear that if $p(d)$ satisfies Estimate 1, then it also satisfies Estimate 2, because $d + M \leq M'd$ with $M' = M + 1$. If Estimate

2 is true then so is Estimate 3; if Estimate 3 is true then Estimate 4 is also true for all d such that $\frac{d}{(\log d)^2} > M$. Finally, if Estimate 4 is true then so is Estimate 5. All this is visible. But what can you prove? What is true? What is false?

P.P. Not every estimate I listed is true. Estimate 1 is false: let M be given, let d be such that $d \geq M$ and $1 + d$ is composite, $1 + 2d > d + M$, hence $p(d) > d + M$.

Paulo. Estimate 1 to the waste basket.

P.P. We shall throw Estimate 2 also into the waste basket.

Eric. Show that Estimate 2 is false.

P.P. OK, given M, let $d = (M + 1)! + 1$. So $1 + d$ is composite, $1 + 2d = 2(M + 1)! + 3$ is composite, $1 + 3d = 3(M + 1)! + 4$ is composite, etc. until $1 + (M - 1)d = (M - 1)(M + 1)! + M$ which is also composite. So $p(d, 1) \geq 1 + Md$, hence $p(d) \geq p(d, 1) > M$. So Estimate 2 is false.

Paulo. And the other estimates? All false?

P.P. Estimate 3 is undecided. It is a conjecture inspired by a thorough and deep study. Estimate 4 is lovely.

Let us write the numbers from 1 to d^2 in a square array with d rows and d columns

$$\begin{array}{ccccc} 1 & 2 & 3 & \ldots & d \\ d+1 & d+2 & d+3 & \ldots & 2d \\ 2d+1 & 2d+2 & 2d+3 & \ldots & 3d \\ \vdots & \vdots & \vdots & \vdots & \vdots \\ (d-1)d+1 & (d-1)d+2 & (d-1)d+3 & \ldots & d^2 \end{array}$$

Estimate 4 may be phrased as follows: if $1 \leq a < d$ and $\gcd(a, d) = 1$, the column a contains a prime number. As you see, this means that $p(d, a) < d^2$ for all a, hence $p(d) < d^2$.

Paulo. I am convinced that one cannot prove the estimate just by making squares with d^2 numbers. But you could find a counterexample.

P.P. People have made squares of d^2 numbers, with d having up to very large values. Estimate 4 has always been found to be true, but there is no proof that Estimate 4 is true for every $d > 1$.

Finally, something has been proved. Linnik showed that there exists a positive integer L, which is effectively computable and such that, for all sufficiently large d, $p(d) < p^t$. Everybody agrees that the proof is extremely hard. Later, researchers could prove that $p(d) < d^{5.5}$ for all sufficiently large d. More work has to be done, in view of replacing 5.5 by 2.

P.P. Now we stop and meet again to sell primes. I will tell you about a factory of primes.

Eric. And I will listen to your story. I want to become rich. Millions.

Notes About Johann Peter Gustav Lejeune Dirichlet (1805–1859)

The name Lejeune Dirichlet means "*le jeune de Richelet*", or the young man from Richelet. Richelet is a small town in Belgium, close to Liège, where the family originated before settling in Düren, a town not far from Köln in Germany. This was the birthplace of Gustav Lejeune Dirichlet at a time when Düren was in the territory of the French Empire. An exceptional student in all subjects, first at the Gymnasium in Bonn and then at the Jesuit College in Köln, where Dirichlet studied mathematics with Ohm. He entered the University of Paris in 1832 and followed lectures at the Collège de France and the Faculté des Sciences by illustrious mathematicians, such as Poisson, Fourier, Lacroix, Legendre and Laplace.

After graduation Dirichlet spent a few years in Paris, after which he returned to Germany. He first held a position in Wroclaw (then called Breslau, in Germany) for a short duration, then in 1832, moved to a position in Berlin. He had a heavy teaching schedule, between the University of Berlin and the Military Academy. In 1855, he became the successor of Gauss in Göttingen, but his tenure was cut short by his death in 1859. Dirichlet was an extremely important mathematician. His production was not voluminous, but his papers contained novel ideas which until today are of central importance.

In his very first work, Dirichlet showed that the equation $x^5 + y^5 = z^5$ has no solution in non-zero integers x, y and z. He first considered only one of two possible cases; independently, Legendre gave a different and full proof of the theorem. Dirichlet proved the remaining case. This paper was published in 1828.

In 1832, Dirichlet proved Fermat's Last Theorem for the exponent 14. He also published a paper on the biquadratic reciprocity law. These results brought the young Dirichlet to the attention of mathematicians as a rising star. These papers were surpassed in importance by his papers of 1837 and subsequent years. Dirichlet used very original and powerful analytic methods to prove his famous theorem on primes in arithmetic progressions: if $a \geq 1$, $d \geq 2$ and $\gcd(a,d) = 1$, there are infinitely many primes among the terms of the arithmetic progression a, $a+d$, $a+2d$, $a+3d$, ... Dirichlet studied the class of series, now called *Dirichlet series*, of which the Riemann zeta series is a main example. And he gave a formula for the number of classes of binary quadratic forms. These were spectacular advances over what had been known before.

His theorem of structure of the group of units of algebraic number fields is also of central importance. Outside number theory, Dirichlet's contributions were no less fundamental. They concern the theory of convergence of trigonometric series, the theory of Fourier series, the study of harmonic functions with boundary conditions giving rise to the problem known today as Dirichlet's problem. Dirichlet also gave the solution of the differential equation in an important hydrodynamics problem, and he studied the stability of the solar system and also contributed in other areas. Dirichlet was honored by academies and scientific institutions during his life and earned the admiration and respect of posterity. A note about his personal life — he was married to Rebecca Mendelssohn, one of the two sisters of the famous composer.

26

Selling Primes

P.P. Today we are selling primes. I will tell you one of my favorite stories, which may inspire you to become rich.

Paulo. Is your story true?

P.P. True or not true — but it will be funny.

The Story. I am a big shot in a factory that produces primes. I will relate to you an interesting dialogue with a buyer, coming from an exotic country.

Buyer: I wish to buy some primes.

I (generously): I can give to you, free of charge, many primes: 2, 3, 5, 7, 11, 13, 17, 19, ...

Buyer (interrupting my generous offer): Thank you, sir; but I want primes with 100 digits. Do you have these for sale?

I: In this factory we can produce primes as large as you wish. There is, in fact, an old method of Euclid, which you may have heard about. If I have any number n of primes, say $p_1, p_2, \ldots, p_n$, we multiply them and add 1, to get the number $N = p_1 p_2 \ldots p_n + 1$. Either N is a prime or, if it is not a prime, we pick any prime dividing N. In this way, it is easy to see that we get a prime which is different from the ones we mixed. Call it p_{n+1}. If we now mix $p_1, p_2, \ldots, p_n, p_{n+1}$ as I already said, we get still another prime p_{n+2}. Repeating this procedure we get as many primes as we wish and so we are bound to get primes as large as we wish, for sure with at least 100 digits.

Buyer: You are very nice to explain your procedure. Even in my distant country, I have heard about it. It gives primes that may be arbitrarily large. However, I want to buy primes that have exactly 100 digits, no more, no less. Do you have them?

I: Yes. Long ago — at the beginning of the nineteenth century — Bertrand observed that between any number $N > 1$ and its double $2N$, there exists at least one prime number. This experimental observation was confirmed by a rigorous proof by Chebyshev. So I can find the primes p_1, p_2, p_3 where

$$10^{99} < p_1 < 2 \times 10^{99}$$
$$2 \times 10^{99} < p_2 < 4 \times 10^{99}$$
$$4 \times 10^{99} < p_3 < 8 \times 10^{99}.$$

Buyer: This means that you have guaranteed three primes with 100 digits, and perhaps a few more. But I want to buy many primes with 100 digits. How many can you produce?

I: I have never counted how many primes of 100 digits could eventually be produced. I have been told that my competitors in other factories have counted the total number of primes up to 10^{21}. We usually write $\pi(N)$ to denote the number of primes up to the number N. Thus, the count I mentioned has given:

$$\pi(10^8) = 5761455$$
$$\pi(10^9) = 50847534$$
$$\pi(10^p) = 455052511 \quad \pi(10^{12}) = 37607912018$$

Selling Primes

$$\pi(10^{17}) = 2625557157654233$$
$$\pi(10^{21}) = 21127269486018731928$$

Even though all primes up to 10^{21} have not yet been produced by any factory, the count of $\pi(10^{21})$ is exact.

Buyer (a bit astonished): If you cannot — as I understand — know all primes of each large size in stock, how can you operate your factory and guarantee delivery of the merchandise?

I: Your country sells oil, does it not? You can estimate the amount of oil at shallow depths quite accurately, but you cannot measure exactly the entire amount underground. It is about the same with us.

Gauss, one of the foremost scientists, discovered that

$$\pi(N) \Big/ \frac{N}{\log N} \quad \text{is approximating equal to 1}$$

for large values of N. This was confirmed, just over a century ago, by a proof given by Hadamard and de la Vallée Poussin.

Buyer: Do you mean that $\pi(N)$ is approximately equal to $N/\log N$, with a small error?

I: Yes. To be more precise, the relative error, namely the absolute value of the difference $\pi(N) - N/\log N$, divided by $\pi(N)$, tends to 0, as N increases indefinitely.

Buyer: Then because of the error you cannot be very specific in your estimate, unless you estimate the error.

I: Correct (*the buyer is not stupid...*). Chebyshev showed, even before the prime number theorem was proved, that if N is large, then

$$0.9 \frac{N}{\log N} < \pi(N) < 1.1 \frac{N}{\log N}.$$

To count primes with 100 digits:

$$0.9 \frac{10^{99}}{99 \log 10} < \pi(10^{99}) < 1.1 \frac{10^{99}}{99 \log 10}$$

$$0.9 \frac{10^{100}}{100 \log 10} < \pi(10^{100}) < 1.1 \frac{10^{100}}{100 \log 10}.$$

It is easy to estimate the difference $\pi(10^{100}) - \pi(10^{99})$, which gives the number of primes with exactly 100 digits:

$$3.42 \times 10^{97} < \pi(10^{100}) - \pi(10^{99}) < 4.38 \times 10^{97}.$$

Buyer: You are rich; I think you have more primes than we have oil. But I wonder how your factory produces the primes with 100 digits. I have an idea but I'm not sure how efficient my method would be.

(1) Write all the numbers with 100 digits.
(2) Cross out, in succession, all the multiples of 2, of 3, of 5, ..., of each prime p less than 10^{99}. For this purpose, spot the first multiple of p, then cross out every p^{th} number.

What remains are the primes between 10^{99} and 10^{100}, that is, the primes with exactly 100 digits.

I: This procedure is correct and was already discovered by Eratosthenes (in the 3rd century B.C.). In fact, you may stop when you have crossed out the multiples of all the primes less than 10^{50}.

However, this method of production is too slow. This explains why the archeologists never found a factory of primes among the Greek ruins, but just temples to Apollon, statues of Aphrodite (known as Venus, since the time of Romans), and other ugly remains, which bear witness to a high degree of decadence.

Even with computers, this process is too slow to be practical. Think of a computer that writes 10^6 digits per second.

Selling Primes

- There are $10^{100} - 10^{99} = 10^{99} \times 9$ numbers with 100 digits.
- These numbers have a total of $10^{101} \times 9$ digits.
- One needs $10^{95} \times 9$ seconds to write these numbers, about 1.5×10^{94} minutes, or about 25×10^{92} hours, so more than 10^{41} days. That is of the order of 3×10^{88} years; 3×10^{86} centuries!

And after writing the numbers (if there is still an After ...) there is much more to be done!

Before the buyer could complain, I added:

I: There are shortcuts, but even then the method would still be too slow. Instead of trying to list the primes with 100 digits, our factory uses fast algorithms to produce enough primes to cover our orders.

Buyer: I am amazed. I never thought how important it was to have a fast method. Can you tell me the procedure used in your factory? I am really curious.

Yes, this buyer was being too nosy. Now, I am convinced that he was a spy.

I: When you buy a Mercedes, you don't ask how it was built. You choose your favorite color: pink, purple, or green with orange dots. You drive it and you are happy because everyone else is envious of you. Our factory will deliver the primes you order and we will do better than Mercedes. We support our product with a lifetime guarantee. Goodbye, sir.

He may have understood:

Buyer: Good buy, sir ...

After the story, Papa Paulo said:

P.P. I hope after this dialogue with the spy-buyer you became curious to know about our fast procedure to produce large primes. I shall tell you some of our most cherished secrets. But swear that you will not share my revelations with anyone else.

Paulo and Eric said nothing, so I decided to just give them a sketch of the production method.

P.P. In the beginning of the last century, Pocklington — who liked to play with numbers and definitely not liked thinking of producing primes — proved the following theorem:

Theorem. *Let p be an odd prime, let k be a positive integer not a multiple of p and such that $1 \leq k < 2(p+1)$. Let $N = 2kp + 1$. Assume that there exists an integer a, $1 \leq a < N$, such that $a^{kp} \equiv -1 \pmod{N}$ and $\gcd(a^k + 1, N) = 1$. Then N is a prime.*

This is what I will prove: If q is a prime factor of N such that $1 < q \leq N$ then $q > \sqrt{N}$.

If this has been shown, then N cannot have two prime factors (equal or distinct), because their product would be bigger than N. So N is a prime.

Paulo. OK, I agree with your strategy.

P.P. I begin.

Proof: Let q be a prime factor of N so $q \neq 2$. Then $a \equiv -1 \pmod{q}$, hence $a^{2kp} \equiv 1 \pmod{q}$, that is, $a^{N-1} \equiv 1 \pmod{q}$. Let $e \geq 1$ be the order of a modulo q. From Fermat's little theorem, $a^{q-1} \equiv 1 \pmod{q}$. Hence e divides $q - 1$ and also e divides $2kp = N - 1$. I note that $a^k \not\equiv 1 \pmod{q}$.

Otherwise $a^k \equiv 1 \pmod{q}$, so $a^{kp} \equiv 1 \pmod{q}$, hence q divides 2, so $q = 2$ and N would be even — which is not true. So I proved that $a^k \not\equiv 1 \pmod{q}$. By assumption $\gcd(a^k + 1, N) = 1$, so q does not divide $a^k + 1$, that is $a^k \not\equiv -1 \pmod{q}$. Combining with $a^k \not\equiv 1 \pmod{q}$ then $a^{2k} \not\equiv 1 \pmod{q}$. Therefore e does not divide $2k = \frac{N-1}{p}$. This is the situation: e divides $N - 1 = p \times \frac{N-1}{p}$.

By the unique factorization theorem, p divides e, which in turn divides $q-1$. So p divides $q-1$, hence $2p$ divides $q-1$, and $2p \leq q-1$, therefore $2p+1 \leq q$. To conclude, $N = 2kp+1 \leq 2p \times 2(p+1)+1 = 4p^2 + 4p + 1 = (2p+1)^2 \leq q^2$. Finally, $\sqrt{N} \leq q$. q.e.d.

Paulo. After all you taught us, Papa Paulo, this proof was pretty easy to follow, but less easy to discover and a bit contorted.

Eric. I agree about the proof. But, in my view there is a question about the statement. How do you find the integer a such that $a^{kp} \equiv -1 \pmod{N}$ and $\gcd(a^k + 1, N) = 1$?

P.P. We wish to obtain a prime $N = 2kp + 1$. This is what we do: We take $a = 2$ and k as it was said. If $2^{kp} \equiv -1 \pmod{N}$ and $\gcd(2^k + 1, N) = 1$, the theorem tells that $N = 2kp + 1$ is a prime. We smile. We are lucky and happy. If the pair of numbers $a = 2$ and k, as it was given, do not satisfy the required conditions, we choose a number k_1 to replace k, such that k_1 is not a multiple of p and $k_1 < 2(p + 1)$. If the required conditions are satisfied by $a = 2$ and k_1, we are happy and $N_1 = 2k_1p + 1$ is a prime. If k_1 does not serve our purpose, we choose another number k_2. We keep repeating this a few more times. If we are still not lucky, we become angry with $a = 2$ and replace it by $a = 3$. Again, we try values of k, until we are lucky. If we still have no luck, we replace $a = 3$ by $a = 5$ and repeat the same trials.

Eric interrupted and said:

Eric. How often have you to do these trials with $a = 2$, then 3, then 5, etc. and successive values of k?

P.P. It so happens we know it by experience that if $a = 2, 3$ or 5, there always exists k, not a multiple of p, such that the required conditions are verified.

Eric. You said "by experience". Is this one of the problems that even the great specialists in analytic number theory have not yet understood?

P.P. The justification of what we observed by experience is a very difficult problem. There are only very limited results which are based on outstanding conjectures and heuristics.

Paulo. You used the new word "heuristics". What does it mean?

P.P. Heuristic results are the ones formulated after trial and error, often based on numerical calculations.

Papa Paulo continued:

P.P. You don't have to worry when you are producing primes. You have such a freedom of choices for a and k, that success is guaranteed.

Papa Paulo could feel Eric and Paulo's reservations. They were disappointed that there was no sure, deterministic way to go about finding primes. But Papa Paulo wanted to convince them that the method was fast and easy to use. He said:

P.P. Using this method let us write a prime that has 100 digits.

We begin with the prime p_1. We find a_1, k_1 as indicated and obtain the prime $p_2 = 2k_1 p_1 + 1$. Then we continue, keeping track of the number of digits of the primes $p_1, p_2, \ldots$ until we reach a prime with exactly 100 digits.

Here is an example.

$p_1 = 2333$
$p_2 = 9336667$
$p_3 = 174347410924471$
$p_4 = 60794039392135489148308051219$
$\qquad k_1 = 2001 \qquad a_1 = 2$
$\qquad k_2 = 9336705 \qquad a_2 = 3$
$\qquad k_3 = 174347410924479 \qquad a_3 = 2$
$\qquad k_4 = 60794039392135489148308051256 \qquad a_4 = 3$
$\qquad k_5 = 5000000000000000000000000000000000000137 \qquad a_5 = 2$
$p_5 = 7391830451225043189805749502951935872607026673729850562129$
$p_6 = 73918304512250431898057495029519358726070266737298505621292035273486223963006775363808830429094325308601979054023347$

Paulo said:

Paulo. I will try this process and get another prime with 100 digits. Perhaps I'll be hired by the factory of your story.

Eric. Is there any other fast way to find big primes? Or to recognize if a big number is prime?

Selling Primes

P.P. Yes, there is something I wish to explain. Suppose you write — or someone gives to you — a very large number N and you want to know if N is prime.

Eric. You already told us that there are many methods to decide if N is prime or composite. I don't know all these methods because you said that they would be too technical to explain, but I remember a little bit.

First you check if N does or does not have any small prime factor, say less than 1000. If it does, N is composite. If it does not, you control whether N is of a special shape. If it is, you may apply powerful primality tests, which work even with very large numbers. In this way, the largest Mersenne primes were discovered; Fermat numbers were also tested. You also said that if N is not of special form, you can use tests which are only practical for numbers not having too many digits.

P.P. Assume that N is not of special form and that it is so big that the current methods are not practical to decide if N is prime.

Eric. Then we are in big trouble. What can be done?

P.P. In this situation, mathematicians have invented a bizarre way of attacking the number N. To begin, I will explain a property which must be satisfied if N is a prime.

Assume that $N > 2$ and N is odd. Let $N - 1 = 2^s d$ where $s \geq 1$ and d is odd. If N is prime, for each a such that $1 < a < N$ (hence automatically $\gcd(a, N) = 1$) the following property is satisfied:

(a^*) Either $a^d \equiv 1 \pmod{N}$ or there exists r, $0 \leq r < s$ such that $a^{2^r d} \equiv -1 \pmod{N}$.

Paulo. This must be easy to show because N is prime, so there exists a primitive root g modulo N.

P.P. Yes, this is how we start.

Proof: Given a, there exists t, $0 \leq t \leq N - 2$, such that $a \equiv g^t \pmod{N}$. We write $t = 2^u e$ where $u \geq 0$ and e is odd. If $u \geq s$, then $a^d \equiv g^{td} \equiv g^{2^u e d} \equiv \left(g^{2^s d}\right)^{2^{u-s} e} \equiv 1 \pmod{N}$. If $0 \leq u < s$, then $a^{2^{s-1-u} d} \equiv g^{2^{s-1-u} td} \equiv g^{2^{s-1-u} \times 2^u e d} \equiv \left(g^{2^{s-1} d}\right)^e$.

Now we observe that $g^{2^s d} \equiv g^{N-1} \equiv 1 \pmod{N}$ and $g^{2^{s-1}d} \not\equiv 1 \pmod{N}$, because g is a primitive root modulo N.

But $\left(g^{2^{s-1}d}\right)^2 \equiv 1 \pmod{N}$, hence $g^{2^{s-1}d} \equiv -1 \pmod{N}$. We deduce that $\left(g^{2^{s-1}d}\right)^e \equiv (-1)^e \equiv -1 \pmod{N}$, because e is odd. The proof is concluded. q.e.d.

Eric. OK. All prime numbers greater than 2 satisfy the property (a^*) for all possible integers a. This is what happens when N is a prime. But this is exactly what we want to know: If N is a prime. Oh! I know what you are going to say, Papa Paulo. The converse is true: if N is odd greater than 2, if property (a^*) is satisfied, for all integers a, $1 < a < N$ with $\gcd(a, N) = 1$, then N is a prime. This is neat.

Paulo. So, so, so! You can find if N is a prime in such a simple way, checking a few congruences. That is great news.

P.P. The converse is true, but it is not such great news. You would have too many bases a to check property (a^*).

Eric. But is it possible to have a composite number N and one base a, $2 \leq a < N$, with $\gcd(a, N) = 1$ and such that property (a^*) is satisfied?

P.P. It is indeed possible: $2047 = 23 \times 89$ satisfies property (2^*). Because of this and many more examples, mathematicians decided to study composite numbers satisfying property (a^*) for some a. These numbers N are composite, but share with the prime numbers the strong property (a^*). For this reason we say that N is a *strong pseudoprime for the base* a, if N is composite and satisfies property (a^*). I just want to say N is a $spsp(a)$.

Paulo. You say that 2047 is a $spsp(2)$. But are there many examples?

P.P. It has been proved that for every $a > 1$ there exist infinitely many $spsp(a)$.

Papa Paulo continued:

P.P. The important result in connection with strong pseudoprimes is the following theorem, which is tricky to prove:

Theorem. *If $N > 2$ is odd and composite, there are at most $\frac{1}{4}\varphi(N)$ bases a, $2 \leq a < N$, with $\gcd(a, N) = 1$, such that N is a $spsp(a)$.*

So one can say that the probability for a composite number to be a $spsp(a)$ is at most $\frac{1}{4}$.

Paulo. Why is this important?

P.P. You'll be surprised. Let b be, like a, such that $1 < b < N$, $\gcd(b, N) = 1$. Since there is no reason to believe the contrary the properties (a^*) and (b^*) may be considered to be unrelated. The probability for (a^*) to hold and respectively for (b^*) to hold may be thought as being independent. So the probability that the composite number N be simultaneously a $spsp(a)$ and a $spsp(b)$ is $\frac{1}{4} \times \frac{1}{4} = \frac{1}{16}$.

Now we are ready to listen to the marketing procedures of the factory. This is what they do with the number N.

(1) To pick $k = 30$ numbers a, such that $1 < a < N$ and $\gcd(a, N) = 1$.
(2) To find if N satisfies condition (a^*) for each one of the chosen numbers a. If there exists a such that N does not satisfy (a^*), declare that N is composite.
(3) Assume that N satisfies condition (a^*) for each one of the 30 numbers a chosen. The probability that N be composite and satisfy (a^*) for 30 values of a, is at most $(\frac{1}{4})^{30} = \frac{1}{2^{60}} < \frac{1}{10^{18}}$. It is extremely small. In this case, the manager of the factory gave the following instructions:

If the customer is someone in a scientific field, declare that N is a probable prime. If the customer is in business, declare that N is prime. And the certification is given with total guarantee!

The manager continued, "We may sell the number N as if it were a prime, even with a 'money back guarantee', because the probability that we are selling a composite number is only 1 in every 1 000 000 000 000 000 000 sales! This is a better guarantee than anyone can get in any deal. We are sure that our company will not be bankrupt and will continue to support generously my trips to advertise our products — all complemented by lavish dinners and the finest wines, to help convince our customers that primes are the way of life."

After this marvelous scientific-industrial tale, Papa Paulo said:

P.P. We shall still meet before vacation in order to discuss mysteries. To be in the right atmosphere we shall begin when the 12 midnight bells toll. In the background there will be music.

Paulo. No Beatles, or Viennese waltzes, I suppose.

P.P. "Bluebeard's Castle" composed by Bela Bartok, "Night on Bald Mountain" by Modest Mussorgsky, "Danse Macabre" by Camille Saint-Saëns, "Dance of the Dead" by Franz Liszt, "The Isle of the Dead" by Sergei Rachmaninoff.

Papa Paulo was happy with what was forthcoming. Nothing attracts more than fear and mystery.

27

The Great Prime Mysteries

Eric. We are here to listen to mysteries of prime numbers. I expect that anyone who will explain one mystery — just one is enough — will become famous.

P.P. You are right Eric. An American MM (multi-millionaire) liked mathematics when he was a boy. He wanted to devote his life to the study of his favorite topic, which was in fact, his passion. But his Dad — a capital "D" Dad whom the young man respected very much — said: "Sonny, do not study mathematics, you will be poor. Instead, follow in my footsteps and develop the family's factories. One day — it will not be long — you will be an MM, and you'll be able to undertake your mathematics studies." Sonny followed Dad's advice, became very rich, studied mathematics and ...

Eric. He solved some mystery of prime numbers and also became famous.

P.P. Not so. He founded an International Institute in 2000, at a gala mathematical meeting at one of the most prestigious European institutes. There, the MM announced he was offering a prize of 1 million US dollars to anyone who would solve one of the seven outstanding mathematical problems at the end of the 20th century. Among the problems was the proof of the Riemann Hypothesis.

Eric. I may want to try it. Remind me about it and please tell us what is already known.

The Riemann Hypothesis

P.P. The Riemann zeta function (which I abbreviate RZF) $\zeta(s)$ is defined and analytic for all complex numbers s, except for $s = 1$, where it has a pole of order 1. For all s with real part $\mathrm{Re}(s) > 1$, $\zeta(s) = \sum_{n=1}^{\infty} \frac{1}{n^s}$. The zeros of the RZF are $-1, -2, -3, \ldots$ and also the non-trivial zeros s with $0 < \mathrm{Re}(s) < 1$ (the critical strip). Due to a symmetry property, the vertical line of the points s with $\mathrm{Re}(s) = \frac{1}{2}$ (called the *critical line*) plays an important role. The Riemann Hypothesis (RH) states: The non-trivial zeros of the RZF are on the critical line.

By another symmetry property, it is equivalent to say that the non-trivial zeros s with the imaginary part $\mathrm{Im}(s) > 0$ are on the critical line. It has been shown that for every $t > 0$ there exists at most finitely many non-trivial zeros s of the RZF such that $\mathrm{Im}(s) = t$. This allows us to write these zeros as a sequence of complex numbers $s_1, s_2, s_3, \ldots$, with $\mathrm{Im}(s_n) \leq \mathrm{Im}(s_{n+1})$ and if $\mathrm{Im}(s_n) = \mathrm{Im}(s_{n+1})$ then $\mathrm{Re}(s_n) < \mathrm{Re}(s_{n+1})$.

The calculation of the non-trivial zeros of the RZF is not an easy matter, but it is now well mastered. It has been shown that all zeros s_n, with $n \leq 10^{11}$ (100 billion) lie on the critical line. For example, $s_{10^{11}} = \frac{1}{2} + i \times 29538618432236$. This is positive evidence pointing to (but not a guarantee of) the truth of the RH.

It has also been shown with a theoretical proof (that is, not computational) that 2/5 of the non-trivial zeros of the RZF are on the critical line. The proof is, of course, delicate and may no doubt be somewhat refined to give a higher proportion. But no one anticipates that this approach will lead to a proof of the RH.

Eric. Is this all that exists in the direction of a proof? I understand very little, but to get the zeros on a line ... well, let me tell you something. I can get a straight line by folding paper. For the RH a symmetry has to be important.

Eric continued:

Eric. You could also try to prove that all non-trivial zeros of the RZF are in a strip around the critical line but narrower than the critical strip, say the strip of all $s + it$ where $\frac{1}{2} - r \leq \text{Re}(s) \leq \frac{1}{2} + r$ where r is a real number such that $0 < r < \frac{1}{2}$. After that, try to make $r = 0$.

P.P. I will address your remark first, Eric. Believe me, mathematicians have tried hard to find a strip without zeros of the RZF, of the form: all s such that $1 - r < \text{Re}(s) < 1$ where r is a real number such that $0 < r < \frac{1}{2}$. No success as yet. It is believed that this would be of the same order of difficulty as to prove the RH; of course, this is pure speculation. Mathematicians have sharp minds and are not afraid of mental contortions. They are searching for (hold your breath) a normed space with operators having eigenvalues which are exactly the non-trivial zeros of the RZF. Then to deduce from certain symmetry properties which the operators must have, that the zeros are on the critical line. I cannot explain this to anyone — I would have to first do it for myself. It would be enormously difficult to reconcile the arithmetical content of the RZF with its analytic properties. A respected mathematician went all around presenting his "proof" of the RH based on such a method. An error, which could not be corrected, was found and brought disrepute to this previously reputable mathematician.

Paulo. Why do mathematicians believe that the RH is true?

P.P. The evidence I have indicated earlier is pointing to the truth. Many desirable properties of prime numbers can, at present, only be proved assuming that the RH is true. And the RH is "pretty". People in mathematics or physics like symmetry, hate chaos. To have the zeros on a line is "nicer" than to have them in places which cannot be described in a reasonable way. But there are also the ones who do not believe that the RH is true.

Eric. Papa Paulo, what is your opinion?

P.P. My opinion is that we have already discussed too much about the RH. And there are other mysteries of prime numbers.

Twin Primes

P.P. It is now the time for another great mystery. It concerns *twin primes*.

Paulo, who really likes babies, said with tenderness:

Paulo. Twin primes must be like twin babies. I love them; they play together, they smile at each other. I like them twice as much as one baby alone.

Eric, who likes babies even more than Paulo, said:

Eric. I like triplets, quadruplets and quintuplets, even more. It is a delight when I see them, all hanging close together. So rare.

These interventions set the stage for the discussion. I had to warn them.

P.P. Don't get excited. It is not human twins that we shall discuss. It is about twin primes like 3, 5 or 5, 7 or 11, 13, etc. These are pairs of primes with difference equal to 2, so they must be consecutive primes.

Consulting tables of primes it becomes apparent that the pairs of twin primes, even though somewhat scarce, remain recurring unabated. So it is natural to formulate the...

Twin Prime Conjecture. *There exist infinitely many pairs of twin primes.*

Eric. For me this is true. I am sure that many people must have looked at extended tables of primes and kept finding twin primes. Why should this stop abruptly?

P.P. This is exactly the right way of thinking. But it is not a proof.

Eric. I always ask you the same thing, Papa Paulo. Tell us all what is known about twin primes.

Paulo immediately asked:

Paulo. To show that there are infinitely many pairs of twin primes — to solve the mystery — why do people not adapt Euclid's proof that there exist infinitely many primes?

P.P. I answer you, Paulo. You may try the following. If you already have a bunch of twin primes, you take the double of the product of the bigger primes in each pair of twin primes, then you add 1 to get a number N. You would like to know that N has a factor which is the bigger prime of a pair of twin primes. This is false. Take for example the pairs 3, 5 and 5, 7, we obtain $N = 2 \times 5 \times 7 + 1 = 71$, which does not verify what is required. No method of this kind has ever been found.

Paulo insisted:

Paulo. Then look at Euler's proof. I remember that the sum $\sum \frac{1}{p}$ (for all possible primes) is infinite. So there must exist infinitely many primes. Listen to my idea: consider all the pairs of twin primes p, $p+2$ and make the sum $B = \sum(\frac{1}{p} + \frac{1}{p+2})$, then show that this sum is infinite.

P.P. If you were Norwegian, I would call you Herre Brun. This is exactly what Brun did.

Paulo. So there is no more mystery.

P.P. There is a mystery. The sum B is finite, actually very small: $B = 1.902\ldots$

Paulo. I see trouble, because $1 + \frac{1}{2} + \frac{1}{4} + \frac{1}{8} + \ldots$ has infinitely many summands yet a sum equal to 2.

So the fact that the sum B is finite does not eliminate the possibility that there are infinitely many pairs of twin primes.

P.P. Yes, it just reflects the fact that the pairs of twin primes are far enough apart.

Now I answer Eric's question. First the notation: for every positive real number x, $\pi_2(x)$ denotes the number of pairs of twin primes p, $p+2$, such that $p+2 \leq x$. For example, $\pi_2(3) = \pi_2(4) = 0$, $\pi_2(5) = \pi_2(6) = 1$, $\pi_2(7) = 2$, and so on. We like to have estimates for $\pi_2(x)$. As I said, by the fact that the sum $\sum(\frac{1}{p} + \frac{1}{p+2})$ is finite, we feel that the pairs of twin primes tend to be far apart from each other. So there cannot be too many pairs p, $p+2$ of primes, such that $p+2 \leq x$. This leads to the estimate $\pi_2(x) < \frac{2C_2 x}{(\log x)^2}$.

C_2 is a positive real number, independent of x, which is called the *twin prime constant*. Its approximate value is $C_2 = 0.66\ldots$. The quantitative form of the twin prime conjecture was formulated on a heuristic basis. It is expressed as $\pi_2(x) \sim \frac{2C_2 x}{(\log x)^2}$, so it resembles the Prime Number Theorem.

There have been few theorems, but many calculations about twin primes. For example, it has been proved that for every $m > 1$ there exist m consecutive primes, none of which belongs to a pair of twin primes. Extensive calculations have detected very large pairs of twin primes like $33218925 \times 2^{169690} \pm 1$. The exact value of $\pi_2(x)$ is known for large values of x; for example $\pi_2(10^{15}) = 1177209242304$.

Paulo. Can I compare this with $\pi(10^{15})$?

P.P. $\pi(10^{15}) = 29844570422669$.

Paulo. The quotient $\frac{\pi(10^{15})}{\pi_2(10^{15})} = 24.3$ (approximately) is not too far from $\frac{10^{15}}{15 \log 10}$ divided by $\frac{1.3 \times 10^{15}}{(15 \log 10)^2}$ which is equal to $\frac{15 \log 10}{1.32} = 26.1$ (approximately).

Eric listened and said:

Eric. Nothing that you said comes even close to a method suitable to prove the twin prime conjecture.

P.P. Yes, the twin prime conjecture is a big mystery, even though there is full expectation that it should be true.

P.P. Before any further discussion, I will explain what a triplet of primes is. First there is the triplet 3, 5, 7. Let d_3 be the minimal of the differences $q_3 - q_1$ for all sequences $3 < q_1 < q_2 < q_3$, whence each q_i is a prime. Any triple (q_1, q_2, q_3) with $3 < q_1$ and $q_3 - q_1 = d_3$ is called a *prime triplet*. So, q_1, q_2, q_3 are consecutive primes and $d_3 = 6$.

P.P. The definition of a quadruplet of primes is similar. First there is the quadruplet 3, 5, 7, 11. Next, let d_4 be the differences $q_4 - q_1$, for all sequences of primes $3 < q_1 < q_2 < q_3 < q_4$. A quadruplet of primes is a sequence of four primes as above, with $q_4 - q_1 = d_4$; the primes q_1, q_2, q_3, q_4 must be consecutive and $d_4 = 8$. And now you know how you should define *prime quintuplets*.

Eric. I deduce that no one knows if there are infinitely many triplets of primes, or quadriplets of primes, and so on.

Paulo. Are triplets when p, $p + 2$ and $p + 4$ are primes?

P.P. Paulo, if you had thought one minute, you would not have asked this question.

Eric intervened:

Eric. If $p = 3$, then p, $p+2$ and $p+4$ are 3, 5, 7, which is a triplet of primes. If p is greater than 3, then p is not a multiple of 3. If $p \equiv 1 \pmod{3}$, then $p + 2 \equiv 3 \equiv 0 \pmod{3}$, so $p + 2$ is not a prime. If $p \equiv 2 \pmod{3}$, then $p+2 \equiv 4 \equiv 1 \pmod{3}$ and $p+4 \equiv 6 \equiv 0 \pmod{3}$, so $p+4$ is not a prime. We conclude that the primes of a triplet must be p, $p+2$, $p+6$ or p, $p+4$, $p+6$. Both cases are possible: 5, 7, 11 is of type 2, 6 and 7, 11, 13 is of type 4, 6.

Eric. What about quadruplets of primes? Quintuplets of primes, etc.?

P.P. We have the special quadruplet 3, 5, 7, 11. If $3 < q_1$, then we have the quadruplet 5, 7, 11, 13, with difference 8. This is as close as the primes of a quadruplet can be. Quadruplets are always of type p, $p+2$, $p+6$, $p+8$, because it cannot contain three primes forming the sequence p, $p+2$, $p+4$, or the sequence $p+4$, $p+6$, $p+8$.

Eric. Can you tell us what are the possible types of quintuplets of primes?

P.P. There is the special quintuplet 3, 5, 7, 11, 13 with difference 10. If $5 \leq q_1$ then there are two types of quintuplets: p, $p+2$, $p+6$, $p+8$, $p+12$ and p, $p+4$, $p+6$, $p+10$, $p+12$. Both types are possible: 5, 7, 11, 13, 17, respectively 7, 11, 13, 17, 19. Other types are not possible; I leave the task to show it to you.

Eric. Examples, examples, EXAMPLES!

Theories need examples and if no theorems are proved conjectures may be formulated, that is, we get mysteries to solve.

Papa Paulo stated:

The Prime k-tuplet Conjecture. *Let $k \geq 2$. If there exists a prime k-tuplet $p < p + b_1 < p + b_2 < \cdots < p + b_{k-1}$ with $p > k$, then there exists infinitely many prime k-tuplets of the same type $b_1, b_2, \ldots, b_{k-1}$.*

Eric. Papa Paulo, your conjecture is so embracing that it contains (for $k = 2$) the twin prime conjecture. So its proof is now out of reach.

P.P. If one day the twin prime conjecture is proved to be true, then perhaps a proof of the truth of the prime k-tuplet conjecture will be found. This is what we hope.

The friend of big numbers said:

Eric. If $k = 10^{1000}$, for me it is hard even to find one k-tuplet of primes.

P.P. You have a point. An eventual proof of k-tuplet conjecture will be by theoretical arguments, not computations.

Polignac's Conjecture

P.P. Polignac was a real Prince, with a taste for mathematics and an uninhibited curiosity.

If one asks if there are primes, or even successive primes with difference of 4...

The answer is an immediate yes: 7 and 11.

One can also ask if there are infinitely many pairs of successive primes, or just pairs of primes with difference 4. This is analogue to the twin primes problem and the answer is unknown. One can ask if there are pairs of primes, or pairs of successive primes, with difference equal to 1000, or difference equal to 100 002; even though I cannot give you any example right away, I am convinced that the answer is affirmative and might be found by looking at very extended tables of primes. But it is not the same if one asks, "Is it true that for every even positive integer d there exists a pair of successive primes, or even a pair of primes, with difference equal to d?"

The answer is unknown. Polignac expressed an even stronger belief.

Polignac's Conjecture. *For every positive even integer d there exists infinitely many pairs of successive primes, with difference equal to d.*

Paulo. I believe that to show that every positive integer is the difference of two primes will already take a long time, let alone the full Polignac's conjecture.

P.P. Not much longer than to prove the twin prime conjecture. The method should be similar. No need for discussions. No one has anything intelligent to say about this conjecture.

A Hardy & Littlewood Conjecture

Paulo did wait to say what he observed. He also wanted to formulate a problem. This is what he said:

Paulo. Look. Take $N = 12$, so $\pi(12) = 5$. Take $M \geq 12$, I get for $\pi(M+N) - \pi(M)$:

$$\pi(24) - \pi(12) = 9 - 5 = 4$$
$$\pi(25) - \pi(13) = 9 - 6 = 3$$
$$\pi(26) - \pi(14) = 9 - 6 = 3$$

and so it goes, for example

$$\pi(37) - \pi(25) = 12 - 9 = 3$$

etc. I always get $\pi(M+12) - \pi(M) < 5 = \pi(12)$. So, I state my conjecture:

Conjecture. *For every $N \geq 2$ and for every $M \geq 2$ the number of primes p such that $M < p \leq M + N$ is less than $\pi(N)$. This is also written as $\pi(M+N) - \pi(M) < \pi(N)$.*

P.P. Your experimental observation led to a conjecture which you would want to be called Paulo's conjecture. I am sorry but I have to disappoint you. The same conjecture was first formulated by the pair of distinguished British mathematicians Hardy and Littlewood.

This conjecture is very reasonable. It confirms what we know, namely that primes are rarefying as one advances in the sequence of integers. But to prove the conjecture is quite another matter. There are results which point to the truth of the Hardy & Littlewood conjecture. For example, a hard proof established that $\pi(M+N) - \pi(M) < \frac{2N}{\log N}$.

If you remember, I mentioned earlier another hard result $\frac{N}{\log N} < \pi(N)$, so $\pi(M+N) - \pi(M) < 2\pi(N)$. There are more results in the direction of proving the conjecture, but they are too technical to report.

Eric, wanting to support Paulo's conjecture, said:

Eric. So it is only a question of time before a proof of Paulo's conjecture will be found.

P.P. Here I have to warn you of something strange. While Paulo's conjecture is very plausible, there is another conjecture which is just as plausible as Paulo's.

Paulo. So both conjectures are true. Then what?

P.P. This is the strange point. It has been shown that both conjectures cannot be simultaneously true. This means that either both conjectures are false or one conjecture is true and the other is false. And this is the dilemma. Is Paulo's conjecture true? Or is the other conjecture true? Or are both conjectures false?

Paulo. How can I guess? You did not tell us the other conjecture.

Now came the great revelation.

P.P. Hold your breath. It has been proved that the Hardy & Littlewood conjecture and the Prime k-tuplet conjecture cannot be simultaneously true.

Paulo. What a mess!

Eric. Change subjects, Papa Paulo, I feel uneasy.

Curious George

P.P. Remember Curious George? He was a monkey who touched everything, broke many precious objects, brought disorder all around him, yet everyone liked Curious George. Some mathematicians are like this monkey, curious and not lacking intelligence. They make conjectures which Curious George would also make, had he been a monkey-mathematician. Here are some of Curious George's conjectures.

Curious Oppermann's Conjecture. *For $N \geq 2$:*

$$\pi(N^2 + N) > \pi(N^2) > \pi(N^2 - N).$$

From this, one sees right away that $\pi\big((N+1)^2\big) > \pi\big((N+1)^2 - (N+1)\big) = \pi(N^2 + N) > \pi(N^2)$. So there are at least two primes between successive squares.

Curious Brocard's Conjecture. *If $n \geq 2$ then*

$$\pi(p_{n+1}^2) - \pi(p_n^2) > 4.$$

In other words, there are at least four primes between the squares of two successive primes.

Paulo. Brocard must have been dead when Oppermann formulated his conjecture. Look, take $p_n = N$. If Oppermann's conjecture is true, before reaching $(N+1)^2$ there are already two primes, and before reaching $(N+2)^2$ there are again at least two more primes. So between $p_n^2 = N^2$ and $(N+2)^2$ there are at least four primes. But $p_{n+1}^2 \geq (N+2)^2$, so we get the statement of Brocard's conjecture.

q.e.d.

P.P. All that you said is right, except that Oppermann might have been dead when Brocard stated his conjecture. Surely, he was unaware of Oppermann's conjecture.

Curious Andrica's Conjecture. *If $n \geq 1$, then $\sqrt{p_{n+1}} - \sqrt{p_n} < 1$.*

If Oppermann's conjecture is true, we saw that there are at least two primes between squares of consecutive integers. This in turn implies that Andrica's conjecture is true. Indeed, given p_n, let N

be the largest integer such that $N^2 < p_n$. Then $p_n < (N+1)^2$. But p_{n+1} must also be between N^2 and $(N+1)^2$. Hence $\sqrt{p_{n+1}} - \sqrt{p_n} < (N+1) - N = 1$.

Eric. May I formulate a conjecture? You'll see what follows from my conjecture.

Curious Eric's Conjecture.
$$\lim_{n \to \infty} \left(\sqrt{p_{n+1}} - \sqrt{p_n} \right) = 0.$$

P.P. Maybe someone has already made this conjecture. Which consequence do you want to tell me?

Eric. I may prove this consequence: let M be any positive integer; there exists a positive integer N_0 such that if $N \geq N_0$ between N^2 and $(N+1)^2$ there are at least M primes.

The proof goes as follows.

Proof: Given M, by my conjecture there exists $n_0 \geq 1$ such that if $n \geq n_0$, then $\sqrt{p_{n+1}} - \sqrt{p_n} < \frac{1}{M}$. This is true because I am assuming that $\lim_{n \to \infty} \left(\sqrt{p_{n+1}} - \sqrt{p_n} \right) = 0$. Let N_0 be such that $p_{n_0} < N_0^2$ and let $N \geq N_0$. Let p_n be the largest prime such that $p_n < N^2$, so $p_{n_0} \leq p_n$, that is, $n_0 \leq n$. If $p_{n+M} > (N+1)^2$, then $\sqrt{p_{n+M}} - \sqrt{p_n} > (N+1) - N = 1$. But $\sqrt{p_{n+M}} - \sqrt{p_n} = (\sqrt{p_{n+M}} - \sqrt{p_{n+M-1}}) + (\sqrt{p_{n+M-1}} - \sqrt{p_{n+M-2}}) + \cdots + (\sqrt{p_{n+1}} - \sqrt{p_n}) < \frac{1}{M} + \frac{1}{M} + \cdots + \frac{1}{M} = M \times (\frac{1}{M}) = 1$. We reached a contradiction. So $p_{n+M} < (N+1)^2$, as it was required to prove.   q.e.d.

P.P. Bravo Eric. Your conjecture is strong. The only drawback is that no one knows if it is true.

The Iterated Gaps

P.P. Before your great (great?) grandfather was born, someone was subtracting consecutive prime numbers and listing their gaps. Then he again subtracted consecutive gaps. If the result was positive or zero, no change; if the result was negative, the sign was changed. And he kept repeating the same procedure.

Paulo. Don't explain. Show!

P.P. OK, look at what I get

```
2  3  5  7  11 13 17 19 23 29 31 37 41 43 47 53 59 61...
1  2  2  4  2  4  2  4  6  2  6  4  2  4  6  6  2  ...
1  0  2  2  2  2  2  2  4  4  2  2  2  2  0  4  ...
1  2  0  0  0  0  0  2  0  2  0  0  2  4  ...
1  2  0  0  0  2  2  2  2  0  0  0  2  2  ...
1  2  0  0  2  0  0  0  2  0  2  0  ...
1  2  0  0  2  2  0  2  2  2  2  ...
1  2  0  2  0  2  0  2  0  0  0  ...
1  2  2  2  2  2  2  0  0  ...
1  0  0  0  0  0  2  0  ...
1  0  0  0  0  2  2  ...
1  0  0  0  2  0  ...
1  0  0  0  2  2  ...
1  0  0  2  0  ...
1  0  2  2  ...
1  2  0  ...
1  2  ...
1  ...
```

P.P. What do you see?

Eric. Numbers and numbers, even zeros. But in the first column I only see 1's.

Paulo. Can you do more rows? Maybe the first column will have a number different from 1.

P.P. This was done by many people. Do you know how far the calculations went?

No answer, so Papa Paulo continued:

P.P. The first row contained all the primes less than 10 trillion.

Eric. And, of course, always 1 in the first column, otherwise you would not mention this fact in our discussion about mysteries. Is the first column of 1's a conjecture?

P.P. This is indeed the *Curious Proth Conjecture*.

Eric. More mysteries, Papa Paulo. Any mysteries with addition?

A Letter from Goldbach

P.P. One day, Euler received a letter from his friend Goldbach. It was not quite like this: "Hi, Lenny ... family OK?" It was written in the old style of the 18th century with elegant sentences. It contained the assertion that he (Goldbach) had observed that every integer $n \geq 6$ was the sum of three primes. Of course, this is very easy to verify for small integers: $6 = 2+2+2$, $7 = 2+2+3$, $8 = 3+3+2, \ldots, 13 = 5+5+3$, $14 = 7+5+2, \ldots, 100 = 79+19+2$, etc. Euler liked this observation and wrote back to Goldbach: "It is also true that every even integer $2n \geq 4$ is the sum of two primes." He also said: "I don't know how to prove these assertions, but I know how to prove that if one assertion is true, then so is the other assertion."

Paulo. Was this easy? I mean to prove that each assertion implies the other.

P.P. Easy, and I will do this proof right now.

> **Proof:** (G) shall stand for the *Curious Goldbach Conjecture*: every integer $n \geq 6$ is the sum of three primes.
>
> (E) stands for Euler's statement: every even integer $2n \geq 4$ is the sum of two primes.
>
> Proof that if (G) is true, then (E) is true: Let $2n \geq 4$, then $2n + 2 \geq 6$, hence by the assumption (G) there exist three primes p, p' and p'' such that $2n+2 = p+p'+p''$. Then one of these primes, say $p'' = 2$, hence $2n = p+p'$, which was to be proved.
>
> Proof that if (E) is true, then (G) is true: Let $2n \geq 6$, then $2n - 2 \geq 4$, so by (E) there exist primes p and p' such that $2n - 2 = p+p'$, hence $2n = p+p'+2$. We also have $2n+1 = p+p'+3$. This proves that (G) is true. q.e.d.

Eric. So what! You proved that the assertions (G) and (E) are both true or both false. But you have not decided which is which. I bet that many people have made those calculations and found that for every number considered, the assertion of Goldbach was true. Otherwise there would be no Goldbach mystery.

P.P. It is known that Goldbach was right for all integers up to $8 \times 10^{15} = 8\,000\,000\,000\,000\,000$ (8 quintillions).

Eric. This may indicate that Goldbach was right, or that the computing people were not patient and abandoned the search for a counterexample too soon.

P.P. Non-believers in Goldbach's assertion have also been in action. This is their idea. Assume that there are even integers which are not equal to the sum of two primes. For each $x > 0$ let $G'(x)$ be the number of even integers $2n \leq x$ which are not the sum of two primes. The non-believers wanted to show that $G'(x)$ is not 0; even more, that $G'(x)$ grows — perhaps very slowly — but it grows to infinity. But they could only prove that $G'(x) < x^{1-\frac{1}{25}}$ (this is less them $\frac{x}{2}$ for all large values of x).

Eric. The non-believers cannot be happy with this result, which says that the growth of $G'(x)$ is not more than the growth of $x^{1-\frac{1}{25}}$. And this does not say that $G'(x) \neq 0$. The non-believers had no success.

P.P. There have been theorems which point towards the truth of Goldbach's conjecture.

(a) Every odd integer with at least 7195 digits is the sum of three primes. The proof is technically involved and tells nothing for smaller odd natural numbers.

(b) If an analogue of the Riemann Hypothesis is assumed to be true, it was shown that every odd integer greater than 5 is the sum of three primes.

These two results, strong as they are, still leave work to be done. Can one get (a) also for numbers with less digits, say with at least 50 digits? Then, by long but direct calculations, examine every odd integer less than 10^{50}, one after the other — and hopefully find them to be the sum of three primes. Another task: can we prove (b) without an appeal to a form of the Riemann Hypothesis?

Eric. Good mysteries. But I noted, Papa Paulo, that you only dealt with odd integers.

P.P. For even integers another method came close (if I can say it) to the proof. This is what was shown:

(c) Every even integer is the sum of a prime and another number which is a prime, or the square of a prime, or the product of two distinct primes.

The numbers p^2 or pq (with p, q distinct primes) are called *almost primes*. What is annoying is that there are many almost primes. So the proof of (c) should — but nobody knows how — be improved and ...

Eric. Almost primes thrown in the waste basket, replaced by primes. We need a big waste basket.

Paulo. And what if we cannot prove Goldbach's conjecture? Perhaps there is a fixed number $Q > 0$ such that every odd positive integer is the sum of at most Q primes. You see, Goldbach was saying not only that Q exists, but that $Q = 3$. And I am saying that Q exists. Of course it would be heaven if we show that Q is small. Maybe 3?

P.P. This is exactly what Schnirelmann proved, but instead of Q, we call this number S_0 and today we already know that S_0 is less than or equal to 6.

Eric. But not 3.

P.P. Anyway the method of Schnirelmann is very important in the branch called additive number theory. What I told you is just one of its results.

After a pause, Eric spoke:

Eric. Addition and subtraction are the easiest operations. But when confronted with primes, addition and subtraction give rise to many problems. Why?

P.P. Because the definition of primes involves multiplication and division only. Confrontation with addition and subtraction breeds problems. It is like marrying two persons with very different characters. Explosion or subtle harmony.

Notes About Christian Goldbach (1690–1764)

Born in Königsberg (Germany), Goldbach, while still very young, visited and established direct contact with many of the outstanding mathematicians of his time, like Leibniz, de Moivre, Hermann and several members of the Bernoulli family (Jakob I, Nikolaus I, Nikolaus II, Daniel). In 1718, Goldbach began his association with the Imperial

Academy in St. Petersburg. His career included an appointment as preceptor of the future Tzar Peter II and Tzarina Ana. Goldbach rose to high administrative positions and was thus prevented from pursuing mathematical research as his main activity, having published only a limited number of papers on the theory of series. Beginning around 1730, Goldbach maintained a steady correspondence with Euler for about thirty years. In a letter from 1742, Goldbach made the conjecture for which he is now remembered — every integer greater than 5 is the sum of three primes.

28

Mysteries in Sequences: More But Not All

P.P. We have seen many mysteries about prime numbers. There was the Riemann Hypothesis and the Goldbach conjecture about sums of primes. Various mysteries about differences of primes, like the term prime conjecture, the k-tuplet conjecture, and the Polignac conjecture. Other mysteries were related to the existence of primes in short integral and to the gaps between successive primes. Now we shall consider sequences of primes which are generated by some procedure, first the primes which are values of some polynomial.

The Sophie Germain Primes

P.P. Dirichlet had proved that if a and d are positives integers such that $\gcd(a,d) = 1$, there exist infinitely many integers $n \geq 1$ such that $dn + a$ is a prime. Thus the polynomial $f(X) = dX + a$ has infinitely many values $f(n)$ which are prime numbers. One may ask if there are infinitely many primes p such that $f(p) = dp + a$ is a prime.

Paulo. Why don't you choose special polynomials so that we understand better the question?

P.P. Let $f(X) = X + 1$. Then $p = 2$ is the only prime such that $p + 1$ is a prime. Let $f(X) = X + 2$, now p and $p + 2$ are primes exactly when they are a pair of twin primes, so it is not known if there are infinitely many primes p such that p and $p+2$ are primes. The next simplest situation is obtained with the polynomial $f(X) = 2X + 1$. While studying Fermat's Last Theorem, Sophie Germain considered the primes p such that $2p+1$ is also a prime. These primes are called *Sophie Germain primes*.

Eric. It is very easy to see that among the primes $p < 100$ the following ones are Sophie Germain primes: 2, 3, 5, 11, 21, 29, 41, 53, 73, 83, 89.

Paulo. The main question must be whether there are infinitely many Sophie Germain primes.

P.P. This is indeed the case and the following conjecture is believed to be true, but its proof should be as difficult as the proof of the twin primes conjecture.

Sophie Germain Primes Conjecture. *There exist infinitely many Sophie Germain primes.*

A similar conjecture is made for the primes of the form $dp + a$.

Primes of the Form $m^2 + 1$

P.P. This time we are interested in the prime values of polynomials of degree 2 $f(X) = aX^2 + bX + c$ and we inquire about the values $f(n)$ which are prime. We focus on the special case where $f(X) = X^2 + 1$ and the question becomes: does there exist infinitely many integers $m > 0$ such that $m^3 + 1$ is a prime?

The answer to this question is still unknown, but numerical evidence leads to the conviction that it is true. In the meantime there is the conjecture:

The $m^2 + 1$ Conjecture. *There exist infinitely many integers $m \geq 1$ such that $m^2 + 1$ is a prime number.*

Paulo. Is there a simpler conjecture for the values of polynomials $f(X) = aX^2 + bX_1 + c$?

P.P. We have to avoid that $f(X)$ be a product of two polynomials of degree 1. We have to avoid that there exists a prime p dividing all values $f(n)$, in particular, we must have $\gcd(a,b,c) = 1$. With these primes it is conjectured that there are infinitely many $n \geq 1$ such that $f(n)$ is a prime number.

Eric. I think that for each allowed polynomial $f(X)$ it is easy to find $n \geq 1$ such that $f(n)$ is a prime number.

P.P. Don't be over-optimistic. For example, $f(X) = X^2 + 576239$ is allowed. For every n such that $1 \leq n \leq 401$ the value $f(n)$ is composite, but $402^2 + 576239$ is prime. It is good that the computer was not stopped before reaching 402. The computer was even more patient for $f(X) = X^{12} + 488669$. I will not show it, but this polynomial is not decomposable and $\gcd(f(0), f(1)) = 1$, so it is allowed. For all integers n such that $1 \leq n \leq 616979$, the value $f(n)$ is composite. But $616980^{12} + 488669$ is a prime, which has 70 digits.

Bunyakovskii did not know these calculations; he lived before computers were born. But he had a feeling about prime values of polynomials, so he made a conjecture.

Bunyakovskii Conjecture. *For every allowable polynomial there is an integer $n_0 \geq 1$ such that the value $f(n_0)$ is a prime number.*

Eric. From what you have said for complicated polynomials involving big numbers, it will be difficult to solve the mystery of Bunyakovskii's conjecture. True or false?

P.P. If Bunyakovskii's conjecture is assumed to be true, it is possible to prove the following statement:

For every allowed polynomial $f(X)$ there are infinitely many $n \geq 1$ such that $f(X)$ is a prime number.

Eric. I bet that it is very hard to prove, going from one prime value to infinitely many prime values. And it is surprising.

P.P. Surprising, yes. Hard, no. I am going to prove it now. Remember, we assume that Bunyakovskii's conjecture is true.

Proof: Let $f(X)$ be an allowed polynomial. So there exists $n_0 \geq 1$ such that $f(n_0)$ is a prime. Let $f_1(n) = f(n + n_0)$ for every n.

I shall prove that this polynomial is allowed. Let $f(X) = a_0 X^k + a_1 X^{k-1} + \cdots + a_k$, with $k \geq 1$, $a_0, a_1, \ldots, a_k$ integers and $a_0 \geq 1$. Then $f_1(n) = f(n+n_0) = a_0(n+n_0)^k + a_1(n+n_0)^{k-1} + \cdots + a_k$. Writing explicitly the results of the operations and reorganizing, $f_1(n) = b_0 n^k + b_1 m^{k-1} + \cdots + b_k$, with $b_0, b_1, \ldots, b_k$ integers, $b_0 = a_0 \geq 1$. So $f_1(X)$ is a polynomial of the same degree as $f(X)$. If $f_1(X)$ is decomposable there exist polynomials $g_1(X)$ and $g'_1(X)$ such that $f_1(X) = g_1(X)g'_1(X)$. Let $g(X) = g_1(X - n_0)$ $g'(X) = g'_1(X - n_0)$, so these polynomials are polynomial $f(X) = f_1(X - n_0) = g_1(X - n_0)g'_1(X - n_0) = g(X)g'(X)$. This is impossible, because $f(X)$ is indecomposable. So $f_1(X)$ is indecomposable. If $m > 1$ and m divides all values of $f_1(n)$, then m divides $f(n) = f_1(n - n_0)$ for all n, and this is absurd. All this shows that $f_1(X)$ is an allowed polynomial. From the assumption that Bunyakovskii's conjecture is true, there exists an integer $n_1 \geq 1$ such that $f(n_1 + n_0) = f_1(n_1)$ is a prime. We note that $n_1 + n_0 \geq 1 + n_0 > n_0$. The argument may be repeated and leads to the required conclusion. q.e.d.

P.P. Encouraged ... No, I shouldn't say that. Unaware of what Bunyakovskii had conjectured, Schinzel and Sierpiński made a conjecture involving several allowed polynomials simultaneously.

Conjecture of Schinzel and Sierpiński. *Let $s \geq 1$, let $f_1(X)$, $f_2(X), \ldots, f_s(X)$ be allowed polynomials. Assume that there does not exist any integer $k > 1$ such that k divides all the products $f_1(n)f_2(n) \ldots f_s(n)$ for all integers n. Then there exists $n_0 \geq 1$ such that $f_1(n_0), f_2(n_0), \ldots, f_s(n_0)$ are primes.*

If the conjecture of Schinzel and Sierpiński is assumed to be true, it is possible to see that there exists infinitely many integers $n \geq 1$ such that $f_1(n), f_2(n), \ldots, f_s(n)$ are simutaneously primes. When $s = 1$, the conjecture of Schinzel and Sierpiński is identical with the conjecture of Bunyakovskii. A special case of the S&S conjecture occurs when $f_1(X) = a_1 X + b_1$, $f_2(X) = a_2 X + b_2, \ldots, f_s(X) = a_s X + b_s$ with $a_1 > 0, a_2 > 0, \ldots, a_s > 0$, $b_1 \neq 0, b_2 \neq 0, \ldots, b_s \neq 0$ and $\gcd(a_1, b_1) = 1$, $\gcd(a_2, b_2) = 1, \ldots, \gcd(a_s, b_s) = 1$. This special case had been considered by Dickson, who stated the corresponding conjecture. But both Bunyakovskii and Dickson did not draw consequences of their conjectures. It was a remarkable accomplishment of Schinzel and Sierpiński to have derived numerous consequences of their conjecture.

Paulo. Do I understand that you are going to assume that the S&S conjecture is true (or just the conjecture of Bunyakowskii, or of Dickson is true) and then you will enumerate mysteries whose solutions are consequences of those conjectures?

P.P. This is exactly what I will do. The conjecture of Schinzel and Sierpiński could be called the *Grand Conjecture*. It is so powerful that if it is proved then quite a number of mysteries — not small mysteries as you will see — will be solved.

Eric did not think long before he said:

Eric. If you can solve all sorts of mysteries just by proving the Grand Conjecture, then I don't think that anyone will ever prove this conjecture. So I don't care whether it is true or false. If I am faced with the proof of one mystery, like the twin prime problem, I'll just look at twin primes, not whether the islands of Faroe are north of Iceland.

Paulo. Eric, until you are proved wrong, you are right. But I still find it... what shall I say? — instructive to know that mysteries are sometimes mutually related. So I want to listen to what Papa Paulo has to say.

With this "go ahead" Papa Paulo felt free to continue:

P.P. I will choose some of the mysteries which may be proved assuming the validity of the conjectures of Dickson, Bunyakovskii or Schinzel and Sierpiński. Ready?

The $m^2 + 1$ Mystery. *There are infinitely many positive integers n such that $m^2 + 1$ is a prime.*

This can be proved as a consequence of Bunyakovskii's conjecture. I had already said it.

The Polignac Conjecture and in particular the Twin Prime Conjecture. *For every even integer $2s \geq 2$, there exist infinitely many pairs of consecutive primes p and p' with difference equal to $2s$. In particular, there exist infinitely many pairs of twin primes.*

This can be proved as a consequence of Dickson's conjecture.

The Strong Twin Prime Conjecture. *For every even integer $2m \geq 2$ there are $2m$ consecutive primes which are m pairs of twin primes.*

This can be proved as a consequence of Dickson's conjecture.

The Sophie Germain Conjecture. *There exist infinitely many Sophie Germain primes.*

This can be proved as a consequence of Dickson's conjecture.

Consecutive Primes in Arithmetic Progression. *Let $n > 1$, let d be a multiple of each prime less or equal to n. Then there exist infinitely many primes p such that $p, p+d, p+2d, \ldots, p+(n-1)d$ are consecutive primes in arithmetic progression.*

This can be proved as a consequence of Dickson's conjecture.

Sophie Germain Primes in Arithmetic Progression. *For every $m \geq 3$ there exist infinitely many arithmetic progressions consisting of m Sophie Germain primes.*

This can be proved as a consequence of Dickson's conjecture.

Composite Mersenne Numbers. *There exist infinitely many composite Mersenne numbers.*

This can be proved as a consequence of Dickson's conjecture.

There was silence. Eric and Paulo were listening with incredulity. Papa Paulo paused and said:

P.P. Before I go to the next statement, I want to remind you of facts discussed a while ago. Remember, if p is an odd prime, there exists an integer a such that $1 < a < p$, $a^{p-1} \equiv 1 \pmod{p}$ and $a^h \not\equiv 1 \pmod{p}$ when $1 \leq h < p-1$. The element a was called a primitive root modulo p. I also said that $a \neq p-1$ and a is not a square modulo p.

Paulo. Yes, I remember. I also recall that you did not prove that a primitive root exists for all odd primes p.

Eric. You used it a number of times in various proofs. We agreed that it would be quite good for calculations if the primitive root modulo p was small, say equal to 2.

P.P. Yes, you have a good memory. The fact is that we cannot predict if a prime p has a primitive root equal to 2. This would be advantageous for calculations. So it becomes natural to ask if 2 is a primitive root modulo infinitely many primes.

Paulo. Why do you like 2 so much? 3 is also small.

P.P. Actually the considerations are the same if, instead of 2, one takes any integer $a \neq 0$, $a \neq -1$ and which is not a square. And there is an important conjecture:

Artin's Conjecture. *If a is an integer, $a \neq 0$, $a \neq -1$ and a is not a square, there exist infinitely many primes p such that a is a primitive root modulo p.*

This can be proved (not so easily) as a consequence of Dickson's conjecture.

Now Papa Paulo turned his attention to ancient Pythagoras.

P.P. Do you remember what Pythagoras proved? If a right-angled triangle has sides measuring a, b and c (the hypothenuse), then $a^2 + b^2 = c^2$. He got in big trouble when $a = b = 1$, because then $c = \sqrt{2}$, so c is not an integer. But if $a = 3$ and $b = 4$, then $9 + 16 = 25$, so $c = 5$; also, if $a = 5$ and $b = 12$, then $25 + 144 = 169$, so $c = 13$. A right-angled triangle with sides measuring a, b and c, which are positive integers, is called a *Pythagorean triangle*. You have just seen two examples.

Eric. Are there infinitely many Pythagorean triangles?

P.P. Yes, and they are easy to obtain. This is how we get Pythagorean triangles. Choose any two positive integers m, n such that $m > n$ (and $\gcd(m, n) = 1$, if you like). Take
$$\begin{cases} a = m^2 - n^2 \\ b = 2mn \\ c = m^2 + n^2. \end{cases}$$
Then
$$\begin{aligned} a^2 + b^2 &= (m^2 - n^2)^2 + (2mn)^2 \\ &= m^4 - 2m^2n^2 + n^4 + 4m^2n^2 \\ &= m^4 + 2m^2n^2 + n^4 = (m^2 + n^2)^2 \\ &= c^2. \end{aligned}$$

Eric. OK, very neat. So if $m = 2$, $n = 1$, then $a = 3$, $b = 4$ and $c = 5$. If $m = 3$, $n = 2$, then $a = 5$, $b = 12$ and $c = 13$.

Are there other ways to obtain Pythagorean triangles?

P.P. No, this is the only way. It is not hard to show, but I'll not do it. I want to observe that in a Pythagorean triangle, the side b is even and greater than 2. In the two examples, the other two sides 3, 5, respectively, 5, 13 are prime numbers. To abbreviate, I shall say that a *prime Pythagorean triangle* is one having two sides a, c which are prime numbers.

Eric asked again a question, a more refined one:

Eric. Are there infinitely many prime Pythagorean triangles?

P.P. There is the following conjecture:

Prime Pythagorean Triangles Conjecture. *There exist infinitely many prime Pythagorean triangles.*

This conjecture can be proved as a consequence of the conjecture of Schinzed and Sierpiński. I will prove it for you, just because I am in love with Pythagorean triangles. The proof goes in three steps:

Proof:
(1) Let a and d be distinct positive integers. Then there exist infinitely many pairs of primes p, q such that $ap^2 + (d-a) = dq$.

To prove (1) let $f_1(X) = dX + 1$ and $f_2(X) = adX^2 + 2aX + 1$. These polynomials are allowed, just note that $f_1(0) = 1$, $f_2(0) = 1$ and $(2a)^2 - 4ad = 4a(a-d)$ is not a square, because $a \neq 0$, $d \neq 0$, $a \neq d$. This suffices to show that the polynomial $f_2(X)$ is not decomposable. We also have $f_1(0)f_2(0) = 1$. By the conjecture of Schinzel and Sierpiński, which we are assuming to be true, there exists infinitely many integers $n \geq 1$ such that $dn + 1 = p$ (prime) and $adn^2 + 2an + 1 = q$ (prime); thus we obtain infinitely many pairs of primes p, q. And we have $ap^2 + (d-a) = a(dn+1)^2 + (d-a) = ad^2n^2 + 2adn + d = d(adn^2 + 2an + 1) = dq$.

(2) If a and d are integers such that $d > a > 0$, there exists infinitely many pairs of primes p, q such that $\frac{d}{a} = \frac{p^2-1}{q-1}$.

By (1) there exist infinitely many pairs of primes p, q such that $ap^2 + (d-a) = dq$. Then $a(p^2 - 1) = d(q-1)$, hence $\frac{d}{a} = \frac{p^2-1}{q-1}$.

(3) Last step: Let $d = 2$ and $a = 1$. Then there exist infinitely many pairs of primes p, q such that $\frac{2}{1} = \frac{p^2-1}{q-1}$. It follows that $p^2 + (q-1)^2 = p^2 + q^2 - 2q + 1 = q^2$. Therefore we obtain infinitely many prime Pythagorean triangles. q.e.d.

Square Arrays of Positive Integers

P.P. Once I showed you the square array of n^2 positive integers (with $n \geq 2$):

1	2	3	...	n
$n+1$	$n+2$	$n+3$	...	$2n$
$2n+1$	$2n+2$	$2n+3$	...	$3n$
⋮	⋮	⋮	⋮	⋮
$(n-1)n+1$	$(n-1)n+2$	$(n-1)n+3$	...	n^2

Do you remember what I said?

Paulo's excellent memory sparked his answer:

Paulo. You said that if $1 \leq k < n$ and $\gcd(k, n) = 1$, then the column containing k has at least one prime.

P.P. Yes, that is what I said, but I also stressed that this was a conjecture. To prove it is still a mystery.

Eric, who thrives on mysteries, continued:

Eric. You stated the mystery of columns. My curiosity forces me to ask something. It is just curiosity. I have no idea what it means.

P.P. Go ahead and ask it anyway. Scientific progress comes from curiosity.

Eric. Can one say that each row of the array contains a prime?

P.P. That is what Sierpiński conjectured:

Sierpiński Conjecture. *Every row of the array contains a prime.*

Eric. Why did he make this conjecture?

P.P. If you look at the array, it is clear that the first row contains a prime. It follows from Bertrand's statement, the one which was proved by Chebyshev, that the second row also contains a prime. For the other rows there is trouble. Using the Prime Number Theorem, it is possible to show that if I give you any integer $h \geq 3$, there is a positive integer n_0 which depends on h, such that if $n > n_0$ there is a prime in each one of the first h rows of the array with n^2 integers. As an example, try the array with $n \geq 13$ rows and verify that each of the first 13 rows contains a prime. The conjecture of Sierpiński has never been proved. It is a mystery.

Eric. It might be very hard to prove it, because even using the Prime Number Theorem it has not been possible to prove Sierpiński's conjecture. Are there consequences of the conjecture of Sierpiński?

P.P. You can easily prove the conjecture of Opperman, if the conjecture of Sierpiński is assumed to be true. I show it. To prove that $\pi(N^2) > \pi(N^2 - N)$, we consider the array with N^2 integers. The last two rows are

$$(N-2)N + 1 \quad \ldots \quad (N-1)N$$
$$(N-1)N + 1 \quad \ldots \quad N^2$$

So $\pi(N^2) > \pi((N-1)N) = \pi(N^2 - N)$.

Next, we consider the array with $(N+1)^2$ integers and look at the two rows

$$(N-2)(N+1)+1 \quad \ldots \quad (N-1)(N+1) = N^2-1$$
$$(N-1)(N+1)+1 = N^2 \quad \ldots \quad N^2+N$$

Then $\pi(N^2+N) > \pi(N^2-1) = \pi(N^2)$ — note that N^2 is not a prime. So I proved Opperman's conjecture as a consequence of the assumption that Sierpiński's conjecture is true.

Eric. And I remember that this implies: (1) between the squares of two consecutive integers there are at least two primes; (2) Brocard's conjecture; (3) Andrica's conjecture. Sierpiński's is strong stuff.

Paulo. What else follows from Sierpiński's conjecture?

P.P. This one is a bit more difficult to show: Between the cubes of two consecutive integers there are at least four primes.

Eric. Another unsolved mystery to add to a long list.

Paulo sighed and commented:

Paulo. There are more prime number mysteries than there are theorems. Long live prime number theory.

Primes in Recurring Sequences

P.P. Do you remember? Once we discussed binary recurring sequences.

Paulo. Just vaguely. Papa Paulo, why don't you refresh our memories? I am not taking ginkgo, so sometimes things slip out of my mind.

P.P. No problem. We did not discuss much about these sequences, so I'll start anew. To define a binary recurring sequence of integers, you need two integers P and Q, which are the *parameters* of the sequence to be defined. You don't want $P=0$ or $Q=0$. Then you also need two integers T_0 and T_1, which will be the *initial terms* of the binary recurring sequence. You don't want that $T_0 = T_1 = 0$.

Now we define $T_2 = PT_1 - QT_0$, $T_3 = PT_2 - QT_1, \ldots$ and $T_n = PT_{n-1} - QT_{n-2}$ for all $n \geq 2$. So each term is a fixed combination, using the parameters, of the two preceding terms.

Paulo. I like letters, but I love numbers more, so Papa Paulo, give concrete examples.

Eric. Hey, Paulo, we have already seen examples. Don't you remember the rabbit story? The Fibonacci numbers were defined taking $P = 1$, $Q = -1$, $F_0 = 0$ and $F_1 = 1$, so $F_2 = F_1 + F_0 = 1$, $F_3 = F_2 + F_1 = 2$, etc. You obtain the sequence

0 1 1 2 3 5 8 13 21 34 55 89 144...

And the Lucas numbers were defined again taking $P = 1$, $Q = -1$ and $L_0 = 2$, $L_1 = 1$. Then $L_2 = L_1 + L_0 = 3$, $L_3 = L_2 + L_1 = 4$ etc. We obtain the sequence of Lucas numbers

2 1 3 4 7 11 18 29 47 76 123 199...

The most convenient binary recurring sequences with given parameters P, Q are of two kinds:

First kind: with initial terms   0, 1.

Second kind: with initial terms   2, P.

P.P. The question I want to discuss is so simple. Does every binary recurring sequence have a term which is a prime number? Infinitely many such terms?

Eric wanted to try different examples. He said:

Eric. I take the two sequences of the first and of the second kind with parameters $P = 2$ and $Q = 1$.

First kind: 0   1   2   3   4   5   6...

Second kind: 2   2   2   2   2   2   2...

The first sequence contains all the prime numbers. The second sequence is a silly sequence, which is degenerate.

Eric. My trial was unlucky. Let me hear what you have to say, Papa Paulo.

P.P. Once I told you the formula for the Fibonacci numbers F_n and for the Lucas numbers L_n. I repeat them here, to have them available for immediate use. Let

$$\gamma = \frac{1+\sqrt{5}}{2} \quad \text{and} \quad \delta = \frac{1-\sqrt{5}}{2}.$$

Then for every $n \geq 0$:

$$F_n = \frac{\gamma^n - \delta^n}{\sqrt{5}} \quad \text{and} \quad L_n = \gamma^n + \delta^n.$$

I have even proved these formulas and I gave shortcuts to calculate F_n and L_n, using powers of matrices.

P.P. Let me see if you agree with me. To find if a Fibonacci number F_n is a prime, you have first to determine which indices n might be such that F_n is a prime. Not all indices are possible.

Eric. OK. And secondly?

P.P. You calculate the Fibonacci number F_n under investigation. It is a number with about $\frac{n}{5}$ digits, so these numbers grow fast when n grows. And the numbers F_n are not of a special form, so to find out if a number F_n is prime will usually be very hard.

Eric. And you are going to say that these smart, ingenious courageous people made incredible calculations and found many primes?

P.P. Wait to hear. Let us start from the beginning. First I want to show that if F_n is a prime then $n = 4$ or n is an odd prime. A preparation is needed.

(1) $F_{2n} = F_n L_n$. Easy to prove:

$$F_n L_n = \frac{\gamma^n - \delta^n}{\sqrt{5}} \times (\gamma^n + \delta^n) = \frac{\gamma^{2n} - \delta^{2n}}{\sqrt{5}} = F_{2n}.$$

(2) If $m \geq n$, then
$$F_{m+n} = F_m L_n - (-1)^n F_{m-n}.$$

Indeed,

$$F_m L_n - (-1)^n F_{m-n} = \frac{\gamma^m - \delta^m}{\sqrt{5}} (\gamma^n + \delta^n) - (\gamma\delta)^n \cdot \frac{\gamma^{m-n} - \delta^{m-n}}{\sqrt{5}}$$

$$= \frac{\gamma^{m+n} - \delta^{m+n}}{\sqrt{5}} = F_{m+n}.$$

because $\gamma^m \delta^n - \gamma^n \delta^m - \gamma^m \delta^n + \gamma^n \delta^m = 0$.

I used the fact that $\gamma \delta = -1$.

(3) If $m \geq 1$, for every $k \geq 1$, F_m divides F_{mk}.

This is true for $k = 1$ and $k = 2$ (by (1)). If there exists k such that F_m does not divide F_{mk}, we choose the smallest such k, which must be greater than 2. We have $F_{mk} = F_{m(k-1)+m} = F_{m(k-1)} L_m - (-1)^m F_{m(k-2)}$. But F_m divides $F_{m(k-1)}$ and also $F_{m(k-2)}$, so F_m divides F_{mk}. This is a contradiction and therefore F_m divides F_{mk} for each $k \geq 1$.

(4) If F_n is an odd prime, then either $n = 4$ or n is a prime.

Proof: Clearly $F_4 = 3$ is a prime. If $n \neq 4$ and n is not a prime, we may write $n = mk$ with $2 < m < n$. Then $F_m \neq 1$ and F_m divides F_n, so F_n would not be a prime, a contradiction. q.e.d.

Eric. So you proved what you announced. Thus the only Fibonacci numbers to investigate are those with odd prime index.

P.P. I salute the mathematicians who did the computations that I report now.

There are 31 primes $p < 300\,000$ such that F_p is prime. The largest is $p = 81839$. There are 6 primes $p < 300\,000$ for which F_p is a probable prime. The largest one is $p = 201\,107$.

Remember? To be a probable prime means that the number was subjected to a battery of primality tests which did not reveal that the number is composite.

Paulo. And what is the feeling of mathematicians about prime Fibonacci numbers?

P.P. Why suddenly is there no more prime Fibonacci numbers? So the conjecture is:

Prime Fibonacci Numbers Conjecture. *There exist infinitely many prime Fibonacci numbers.*

But, since the sequence of Fibonacci numbers is very "thin" (the numbers are not crowded like the streets of Shanghai) the conjecture will be very hard to prove. At present, there are no methods which could lead to the proof.

Eric said, with visible joy:

Eric. The mystery of Fibonacci primes will last for a long time. Great. Now I want to hear the mystery of prime Lucas numbers.

P.P. It is just about the same as for the Fibonacci numbers. First I need some preparation.

(1) $L_{3n} = L_n^3 - 3(-1)^n L_n$.

I will prove...

Paulo. Don't. It is like for Fibonacci numbers. You use the formula for L_n, you make no mistake and you get it. What would have been more difficult is to guess the formula.

P.P. You need a third eye for the intuition.

(2) $L_{m+n} = L_m L_n - (-1)^n L_{m-n}$ when $m \geq n$.

This time I do not menace you with a straightforward proof: "Formula and doing".

(3) If $m \geq 1$, for every odd integer $k \geq 1$ L_m divides L_{km}. The proof is like the one for Fibonacci numbers. This time k is odd and we use the relation $L_{mk} = L_{m(k-2)} L_{2m} - (-1)^{2m} L_{m(k-4)}$.

(4) If L_n is a prime, then $n = 0$ or n is a power of 2 (greater than 1) or n is a prime.

Proof: We assume that L_n is a prime $L_n \neq 2$, that is, $n \neq 0$. So we may write $n = 2^e k$ with k odd, $e \geq 0$. If $k = 1$, then $n = 2^e$, but $n = 1$ is excluded, so $e \geq 1$. Now let $k \geq 3$. Then L_{2^e} divides $L_{2^e k} = L_n$, $L_{2^e} < L_n$; but L_n is a prime so $L_{2^e} = 1$, that is, $2^e = 1$, therefore $n = k$. We show that k is a prime. Otherwise $k = hh'$ with $1 < h < n$ and L_h divides L_n, with $L_h \neq L_1 = 1$. This is impossible, so $k = n$ has to be a prime. q.e.d.

Eric. Now you continue as you did for Fibonacci numbers, giving the results of calculations.

P.P. For $n < 260\,000$ there are 39 Lucas numbers L_n which are primes, the largest being $n = 51169$; there are also 12 Lucas numbers L_n which are probable primes, the largest being $n = 202667$.

Eric added:

Eric. The conjecture must be

Prime Lucas Numbers Conjecture. *There are infinitely many prime Lucas numbers.*

Eric. If I understood well, there are many more binary recurring sequences. I do remember that if $P = 3$ and $Q = 2$, the sequence of the first kind consists of the numbers $2^n - 1$ (for $n \geq 0$) and the sequence of the second kind of the numbers $2^n + 1$ (for $n \geq 0$). We have seen that if $2^n - 1$ is a prime, then $n = p$ is a prime and we get the Mersenne number $M_p = 2^p - 1$. And you have also shown that if $2^n + 1$ is a prime, then $n = 2^m$ (with $m \geq 0$) and you get the Fermat number $F_m = 2^{2^m} + 1$.

Eric. For me what counts are mysteries. And I know that there are two more.

Prime Mersenne Numbers Conjecture. *There are infinitely many prime Mersenne numbers.*

Prime Fermat Numbers Conjecture. *There are infinitely many prime Fermat numbers.*

P.P. People agree that the conjecture about prime Mersenne numbers has to be true. But, of course, no one has any clue on how to try to prove it. But what concerns prime Fermat numbers F_m, the only ones known are for $m = 0, 1, 2, 3$ and 4. If a poll were taken, most mathematicians would agree that there are only five prime Fermat numbers, the ones we already know.

Eric. I gather that because of the disagreement about the conjecture on prime Fermat numbers, one should not conjecture that in every binary recurring sequence of the first or second kind there are infinitely many terms which are prime numbers.

Eric. Papa Paulo, do you think that it is unreasonable to make the conjecture that every non-degenerate binary recurring sequence contains at least one prime term?

P.P. Unreasonable, no. Wrong, yes. There are examples of binary recurring sequences, not of the first or second kind, which do not contain — you hear — any prime terms.

Eric. Well, what a surprise. How did anyone think of finding such a sequence?

P.P. You only find what you seek. I give you one example. The parameters are $P = 1$ and $Q = -1$. The initial terms are
$$T_0 = 62\,63\,82\,80\,04\,23\,98\,57$$
$$T_1 = 49\,46\,34\,35\,74\,32\,05\,65\,5$$

Eric. Is this example rare?

P.P. That is the point. Someone finds an example of something interesting and he writes the proof. Somebody else says: "I can find many more examples by studying how the first one was gotten." And he does it.

Eric. Imitator.

P.P. Intelligent imitator. Sometimes it is far from obvious how to do it. In our context, there exists infinitely many pairs of coprime parameters P, Q such that for each one there exist infinitely many pairs of initial terms such that all the terms of the corresponding sequence are composite integers.

Repunits

P.P. The integers 11, 111, 1111, 11111, ... are called *repunits*.
This means that the digits are repeated units. Instead of writing $111\ldots11$ (50 lines the digit 1), we just write $R50$, and so on.

Eric. Are there prime repunits? I take it back. $R2 = 11$ is a prime. Better question: are there other prime repunits?

P.P. Yes. $R19$, $R23$, $R317$ and $R1031$.

Paulo said:

Paulo. I note that if Rn is a prime, then n is also a prime — in all known examples. Can one prove this in general?

P.P. Yes.

Paulo. Let me prove it. I use the formula for repunits:

$$Rmk = \frac{10^{mk} - 1}{9} = \frac{10^{mk} - 1}{10^m - 1} \times \frac{10^m - 1}{9}.$$

Here I am taking $m > 1$ and $k > 1$. I note — you did it so many times:

$$\frac{10^{mk} - 1}{10^m - 1} = 10^{m(k-1)} + 10^{m(k-2)} + \cdots + 10^m + 1,$$

which is an integer greater than 1. Then $1 < Rm < Rmk$ and Rm divides Rmk. The conclusion is that if Rn is a prime, then n is a prime.

P.P. Bravo. I add that the prime repunits which I have mentioned are the only ones with $n < 60\,000$. There are two repunits which are probable primes: $R49081$ and $R86453$. No hope at present to prove that these two numbers are actually primes. The conjecture is:

Prime Repunit Conjecture. *There are infinitely many prime repunits.*

Eric. One more mystery for my collection.

It became time for us to end the discussions about the seemingly inexhaustible collection of mysteries about prime numbers.

Notes About Marie Sophie Germain (1776–1831)

Sophie Germain, born in Paris (France), was not allowed by her parents to study mathematics. Nevertheless, in hiding she read Newton, Euler and studied the notes of Legendre's lectures at the Ecole Polytechnique. Under the name Monsieur Le Blanc, she sent to Legendre a paper on questions relating to his book "*Essai sur la Théorie des Nombres*". Legendre was struck by the paper's originality and insight. When he learned the true identity of the author, he supported and collaborated with Sophie Germain, including many of her results, in a supplement of the second edition of his book. From 1804 to 1809 she corresponded with Gauss on matters related to *"Disquisitiones Arithmeticae"*. Monsieur Le Blanc's ideas received high praise from Gauss.

The true identity of the French correspondent was revealed when Braunschweig was occupied by the French troops and Sophie Germain, through family connections with the French commander, guaranteed that Gauss would not be harmed; the killing of Archimedes by a Roman soldier would not be repeated with Gauss.

The remarkable theorem of Sophie Germain, included in Legendre's book (1828), stated: if p and $2p+1$ are primes, if p does not divide either one of the non-zero integers x, y, z, then $x^p + y^p \neq z^p$. This was the most important result towards a proof of Fermat's Last Theorem until the work of Kummer in the 1850s and later. In 1815, Sophie Germain received a one-kilogram gold medal from the Académie des Sciences for her work on the elasticity of vibrating plates and the mathematical theory explaining the formation of the Chladni figures. The philosophical essay "Considérations générales sur l'état des sciences et lettres" was published posthumously and well received. Sophie Germain's untimely death deprived mathematics of her superior intelligence.

Notes About Viktor Yakovlevich Bunyakovskii (1804–1899)

Bunkyakovskii was born in Bar, Podolskaya (Ukraine). He received his doctorate in 1825 in Paris under the direction of Cauchy. Returning to St. Petersburg, he brought his expertise on Cauchy's theory of residues and probability theory. He taught in several institutions, including the University of St. Petersburg. He was soon admitted to the Academy and became a full member in 1864. Twenty five years before Schwarz, Bunyakovskii proved the important Cauchy–Schwarz inequality. He also worked in number theory, geometry and mechanics.

Notes About Leonard Eugene Dickson (1874–1954)

Texas-born (Independence, Texas, USA) Dickson had a typical higher level American education: an undergraduate degree at the University of

Texas, a doctoral thesis in 1896 at the University of Chicago, followed by post-doctoral periods in Leipzig and Paris. A number of short-term positions preceded his appointment in 1900 to the University of Chicago, where he taught until his retirement in 1939.

Signs of his intense activity were the 55 doctoral students he supervised, the 18 books he published, as well as the large number of research papers he authored. His specialties were the arithmetic of algebras and representation theory, but he also contributed to Waring's problem. Many of his books were written as teaching texts, but *"Algebras and their Arithmetics"* is at research level. Dickson produced a very useful book in three volumes where he recorded in chronological order, almost without omission, all the papers written in various topics of number theory. It was a truly excellent reference book which is greatly admired.

Notes About Wactaw Sierpiński (1882–1969)

Sierpiński was born in Warsaw (Poland). He entered the University of Warsaw in 1900 and studied under the direction of Voronol. His initial work in number theory concerned the equidistribution therom. Sierpiński wrote a large number of papers on elementary number theory. Most of his research was about set theory, including the Continuum Hypothesis and transfinite numbers, subjects which he treated in two noteworthy books published in 1934 (*"The Continuum Hypothesis"*) and 1958 (*"Cardinal and Ordinal Numbers"*). Sierpiński's book *"Theory of Numbers"* dates from 1964; another of his books contains 250 problems in elementary number theory.

Sierpiński taught first at the University of Lviv and then at Warsaw. He was the recipient of numerous honorary degrees and was a foreign member of several academies.

29

The End and the Beginning

The End of the Discussions

Summer was arriving, and after the mysteries, any discussion about primes would be anti-climactic. Paulo could not disagree, but he still wanted to make — what was for him — a very important remark.

Paulo. Among the mathematicians quoted, only one — Sophie Germain — was a woman. Worse, she had to pose as Monsieur Le Blanc in her correspondence with Gauss.

Eric. It was inexcusable, as we all three agree, that such an intelligent woman had to pose as a man.

Paulo. Even today when women make up about 50.2 percent of the population, less than 10% of mathematicians are women. Papa Paulo, do you know the reason for such an aberration?

Papa Paulo thought a while and knew that his answer had to be "right", lest he caused a turmoil which he would not be able to handle. With little conviction and even less originality, he replied:

P.P. Well, we all know that it was determined that women bear children. We all know how women have been oppressed by men in many cultures.

But in his inner heart Papa Paulo had another idea, which he ventured saying:

P.P. Between us, I have another idea, which no one has ever expressed and I want to write a whole book in support of it. You know, invariably my research ideas appear when I am shaving. Women do not shave and this is an insurmountable biological handicap for mathematical research.

Eric, who had traveled more than anyone else, posed a problem as usual:

Eric. Once I visited the Art Museum in Toledo (not Ohio, but in Spain) and there I saw the painting by Ribera (the "Spagnoletto") called "La Mujer Barbuda" ("The Bearded Woman"). Did she ever shave after the portrait was finished? Did she ever have mathematical thoughts? Papa Paulo, investigate this case; it could be relevant to your theory and destroy your book "Hair and Mathematics", and even more, your reputation.

Eric still wanted to make a remark.

Eric. We talked about a large number of mathematicians which you, Papa Paulo, revered as heroes. Yet, not one of them had fought a war or conquered a kingdom.

P.P. This is indeed the right word. For me, heroes and historic figures are much less the kings, presidents, conquerors, generals ... they are worth much less than scientists — including mathematicians — who fight to understand the unknown, the universe we live in. All in the hope of transforming this knowledge into good for mankind. Hope, difficult to turn into reality, but hope, a motor for any progress. Mathematicians are in my view great unsung heroes. And luckily, they are practicing a science which is also an art. Beauty and symmetry married with rigor and irrepressible imagination.

Vacation

Vacations would bring the perfect balance between brain and muscle. The Greek ideal. Papa Paulo made some suggestions.

P.P. Try one of these:
- Bicycling in the Himalaya
- Riding the pororoca on the Amazon river
- Portaging canoes in the Canadian North, from lake to lake
- Hunting jararacas in Mato Grosso
- Scuba diving off Fernando de Noronha island
- Following the path of Michel Strogoff all through Siberia
- How about riding giant turtles in Galapagos?

I could give further programs, which are more interesting than mowing lawns in Kentucky, even though the grass there is blue.

Don't try to learn a language like papiamento, ouolof, tupí-guaraní, swahili, kabyle or...

Impatient, Eric stopped me and said:

Eric. I listened when you talked about prime numbers. Vacation for me is time for doing nothing, like the number zero which is so valuable behind other numbers. The notes you handed me are not well organized and I give you a vacation task: to rewrite all this in a nice way, just as if it were a book. I don't mind being a part of it as the questioning person.

P.P. This is an idea which I will follow.

Following the Advice

A few days later, I called my favorite publisher, of course Marcel Spank, of Gold Springs Publishing Company in New York.

P.P. Here are notes of discussions about prime numbers. Would you consider examining the text and, who knows, publishing it?

Marcel, whom I normally call Marcelino, more out of friendship than because of his short stature, answered:

Marcelino. Why not? People buy Harry Potter books. If our marketing division does a good job, we could sell your book very well.

You are so famous that a million copies would not surprise me. Whatever is inside the covers counts less than the title.

P.P. I know it. To look scientific I used the word "experiments". To make it clear that anyone, even boys and girls may read it, I included "boys" in the title. In support, justified and needed, to feminists, "girls" are also in the title.

Marcelino. Then what is the title?

Papa Paulo, with the air of someone who shows a gem:

P.P. *"Prime Experiments Explained to Boys and Girls"*.

Marcelino. This is horrible. You cannot sell a book with this title. It may give rise to adverse interpretation. Believe in my selling experience.

In his innocence, Papa Paulo said:

P.P. I just want to tell how important it is for young people to experiment with prime numbers, to understand how they are distributed. The young Gauss, at age 15, was able to discover — if not to prove — the fundamental Prime Number Theorem.

Marcelino. In the times we are in, the title you proposed may bring, from ill-intentioned persons, much disagreement.

To this opinion had already concurred my nephew Bob, my niece Myriam, and Connie, my internationally known secretary.

At this point, Marcelino proposed that the book be titled: "The Story of Two Boys in Love with Prime Numbers". He was astonished by my explosive laughter.

P.P. This is exactly the title of the Japanese version of the book, published in Toyko!

Marcelino paused and said:

Marcelino. Come back with a good selling title.

Two Weeks Later

Two weeks of intense thinking. I found a better title for the book. Reading the title, Marcelino spoke:

Marcelino. Oi! Paulo, despite our long collaboration...

I knew it was bad. Marcelino continued:

Marcelino. Gold Springs Publishing Company of New York will not publish such a book. It is totally unacceptable. A book with the aim of explaining everything about prime numbers in a language without secrets, eliminates the mystery of mathematics and it will be detrimental for the sale of our other books which are excellent to generate confusion and call for more books. Besides, you try to be funny — no — sometimes you are, which is worse.

Science has to be serious, mathematics boring. You need to find a publisher who has courage and a good sense of humor.

And despite our long collaboration, Marcelino continued:

Marcelino. It is an irrevocable "no".

More excruciating thinking was needed for the title of a book which few would read. "Prime numbers" had to be in the title (no more "prime experiments"). The title contains the word "problems", which suggests mysteries and even magic. Trying to reach the popularity of the best selling magician, I added the name "Voldemort" to my own name, "Papa Paulo", which I hope, will become famous. The word "trialogue" will make potential readers run to the internet. I confess that I don't even know if this word exists. And I prepared a beautiful cover for my book:

One question remains.
Will I ever find a brave and sympathetic publisher?

Name Index

Bunyakovskii, Viktor Yakovlevich (1804–1899), 309

Carmichael, Robert Daniel (1879–1967), 99

Chebyshev, Pafnuty Lvovich (1821–894), 234

Dase, Johann Martin Zacharias (1824–1861), 206

de la Vallée Poussin, Charles Jean Gustave Nicolas (1866–1962), 235

Dickson, Leonard Eugene (1874–1954), 309

Dirichlet, Johann Peter Gustav Lejeune (1805–1859), 259

Eratosthenes (circa 276 BCE–circa 196 BCE), 20

Erdős, Pál (1913–1996), 237

Euclid of Alexandria (circa 300 BCE–circa 279 BCE), 14

Euler, Leonhard Paul (1707–1783), 69

Fermat, Pierre de (1601–1665), 57

Fibonacci (Leonardo Pisano) (circa 1170–1250), 156

Gauss, Karl Friedrich (1777–1855), 205

Germain, Marie Sophie (1776–1831), 308

Goldbach, Christian (1690–1764), 288

Hadamard, Jacques Salomon (1865–1963), 236

Korselt, Alwin Reinhold (1864–1947), 99

Legendre, Adrien-Marie (1752–1833), 82

Lehmer, Derrick Norman (1867–1938), 100

Lucas, François Edouard Anatole (1842–1891), 157

Mersenne, Marin (1588–1648), 121

Pascal, Blaise (1623–1662), 158

Plato (circa 428 BCE–circa 348 BCE), 15

Pythagoras of Samos (circa 569 BCE–circa 475 BCE), 157

Riemann, Georg Friedrich Bernhard (1826–1866), 234

Selberg, Atle (1917–2007), 237

Sierpiński, Wactaw (1882–1969), 310

Stieltjes, Thomas Jan (1856–1894), 31

Wiles, Andrew (1953–), 96

Wilson, John (1741–1793), 67

Subject Index

Bit operations, 47

Carmichael's Number, 99
Complex number, 214
Complexity of algorithms, 49
Composite number, 34
Coprime integers, 23

Descent method, 24
Divisor (of an integer), 2

Eratosthenes Sieve, 19
Euclid's Division Theorem, 8
Euclid's Theorem (Existence of
  infinitely many primes), 30
Euclid's Theorem for Perfect
  Numbers, 106
Euler's function, 65
Euler's Theorem, 63
Euler's totient function, 63
Even integer, 6

Factorial, 33
Fermat numbers, 94
Fermat's Last Theorem, 57, 58, 96
Fermat's Little Theorem, 55, 61
Finite induction method, 24
Fermat's Theorem for congruences
  modulo a prime, 55

Greatest common divisor, 11

Integers congruent modulo m, 51

Least common multiple, 13
Legendre's symbol, 78

Mersenne numbers, 115
Möbius function, 44
Multiple of an integer, 6

Natural number, 2
Negative integer, 8
Non-quadratic residue, 78

Odd integer, 6
Order of an integer a $modulo$ n, 71

Pairwise relatively prime integers, 56
Perfect numbers, 105
Pigeon-hole Principle, 55
Polynomial time, 89
Positive number, 8
Prime number, 2, 17
Prime root modulo an integer, 72
Primorial number, 36
Proof by reduction to absurd
  (*Reductio ad absurdum*), 22
Pseudoprime in base 2, 98

Quadratic reciprocity law, 79
Quadratic residue, 78
Quotient (of a division), 8

Real numbers, 185
Relatively prime integers, 23
Remainder (of a division), 8
Residue classes, 51

Sieve method, 18

Wilson's Theorem, 67

www.ingramcontent.com/pod-product-compliance
Lightning Source LLC
Chambersburg PA
CBHW070308230426
43664CB00015B/2668